Managing Marine Environments

Richard A. Kenchington

Taylor & Francis

New York • Bristol, PA. • Washington, D.C. • London

USA	Publishing Office:	Taylor & Francis New York Inc. 79 Madison Ave., New York, NY 10016-7892
	Sales Office:	Taylor & Francis Inc. 1900 Frost Road, Bristol, PA 19007-1598
UK		Taylor & Francis Ltd. 4 John St., London WC1N 2ET

Managing Marine Environments

First published 1990
Printed in the United States of America

Library of Congress Cataloging in Publication Data

Kenchington, R.A.
Managing marine environments / Richard Kenchington
p. cm.
Includes bibliographic references and index.
ISBN 0-8448-1635-3
1.Marine parks and reserves—Mangement. 2. Marine Resources conservation. 3. Marine parks and reserves—Management—Case studies. 4. Marine resource conservation—Case studies.
I. Title.
QH91.75.A1K46 1990
333.91'64--dc20 90-11069
CIP

Contents

Acknowledgements iv
Introduction 1

Chapter 1 Historical Background 6
Chapter 2 Defining Management Areas 14
Chapter 3 Nature of Marine Systems Relevant to Management 28
Chapter 4 The Nature of Management Problems 40
Chapter 5 Establishing a Framework for Management 60
Chapter 6 Developing a Management Plan 75
Chapter 7 Implementation of Management 94
Chapter 8 The Great Barrier Reef: A Brief Introduction and History of Research 109
Chapter 9 The Great Barrier Reef Marine Park: The Campaign for Management of the Reef 118
Chapter 10 The Great Barrier Reef Marine Park: The Constitutional and Legal Setting 126
Chapter 11 The Great Barrier Reef Marine Park: Creating an Operational Base 135
Chapter 12 The Great Barrier Reef Marine Park: Procedures Adopted by the Authority 147
Chapter 13 The Great Barrier Reef Marine Park: as a Multiple Management Approach 158
Chapter 14 The Galapagos Marine Resource Reserve 173
Chapter 15 The Republic of Maldives: Marine 184
Chapter 16 International Arrangements 205
Chapter 17 Prospects for Progress in Marine Environment Management 214
Notes 223
Index 235
About the Author 248

Acknowledgements

The support advice and encouragement, criticism and hard work of many people have made this book possible. I owe particular thanks to all involved in the excitement and collaboration which have resulted in the development of methods for planning and management of the Great Barrier Reef Marine Park. These include the members of the Great Barrier Reef Marine Park Authority, my colleagues on the staff of the Authority, the members of the Great Barrier Reef Consultative Committee, the Marine Parks Section of the Queensland National Parks and Wildlife Service and the users and enthusiasts involved in public participation and plan development.

Particular thanks to Graeme Kelleher, Chairman of GBRMPA and to Claudia Baldwin, John Baldwin, Dan Claasen, Wendy Craik, Jon Day, Peter McGinnity, John O'Dwyer, David Perkins, John Wheeler, and Simon Woodley who payed key roles in the development of the planning and management procedures for the reef.

In Ecuador, my thanks to Roque Sevilla, who invited GBRMPA to assist by providing planning advice, and to Fernando Arcos, Fausto Cepeda, Tito Rodriguez and Pepe Villa, the members of the Technical Commission for planning of the for the Galapagos Marine Resources Reserve.

In the Republic of Maldives, my thanks to Mohamed Saeed and to the island and atoll chiefs and administrators who made us welcome and who discussed problems of the reef and island environment.

The existence of this book would not have been possible without the support of the Pugh Foundation for Fellowship at the Marine Policy Center of the Woods Hole Oeanographic Institution. My particular thanks to Jim Broadus, Tundi Agardy and Arthur Gaines.

The final production would not have happened without the help and support of my wife, Carol, the forbearance of my sons Peter and Thomas, the indexing by Helen Penridge and the editorial liaison of Jung Ra and Gary Steinberg.

My thanks to you all.

Introduction

The purpose of this book is to introduce readers to contemporary multi-disciplinary issues of planning and management of marine environments and natural resources. It is not a comprehensive review of marine management nor does it provide detailed coverage of specialist fields such as fisheries or shipping management. It draws heavily on the experience developed in the first 12 years of planning and management of Australia's Great Barrier Reef Marine Park. Most of the other examples are also tropical, but the problems that have had to be addressed in the tropics are little different for higher latitudes.

Experience of the sea has always included mysterious fluctuations in abundance of fished species and occasional unexplained episodes of fish death, whale beaching, of the sea coloured by algae and of the corpses of strange creatures found dead on beaches. The invisibility of most of the phases which occur in the water column and the limited opportunity for direct observation make the role of marine scientists as trained observers particularly important in any public discussion on the management of marine environments.

In the development of marine environment management the marine science research specialist was often the only professional to identify apparent problems in situations where fishermen and other users saw no departure from normal fluctuations of fortune. In the absence of other perspectives the marine scientist was often cast in, or assumed, the sometimes conflicting roles of advocate for marine management and adjudicator of technical aspects of management proposals to an extent far greater than in an equivalent terrestrial situation.

The advocacy of marine scientists has achieved considerable success in persuading nations to proclaim marine and estuarine protected areas. Nevertheless the step between the heady success of proclaiming marine protected areas and the discipline of achieving effective management is a difficult one. It involves progressing from a dramatic political gesture with relatively little commitment of resources to acceptance of a perpetual commitment of significant management resources.

The generic threats to marine environments have been widely identified during the past 20 years. They are pollution, overexploitation of biological resources, and alienation, particularly of estuarine, coastal, and shallow sea habitat. Recently they have become an important item on the global political agenda.[1]

Technically, most of those threats can be reduced or contained with national and international action. Most present less technological challenge than those that have been faced in developing design and standardization of equipment, training and certification procedures, control systems, and communications infrastructure required for international air travel.

Socially and politically, the problems of marine environment management are very difficult. The scale of processes and the transport of impacts in relation to jurisdictional boundaries present the complex challenge of developing effective and acceptable mechanisms for regulating human impact.

The problem is that aquatic environments are of secondary importance to most people who live within and are immediately affected by the problems of terrestrial environments. Until recently, communities have been quite unaware of the concept and the costs of unmanaged or mismanaged marine environments. That awareness is growing, but acceptance of the inevitability of those costs in one form or another has yet to penetrate community economic management. The medium to long-term options are to witness and absorb the social and economic costs of continuing and increasing loss of environmental processes and amenity or to pay the full costs of management-imposed usage and pollution controls in order to maintain environmental processes and resources.

The underlying priority for those concerned for the management of marine environments is to achieve awareness and acceptance by communities and governments that the costs cannot be avoided. Without management they are passed on so that estuarine and coastal communities endure the consequences.

The components of marine environment management are resource allocation, impact minimization and pollution control. They may be applied to achieve control of the impacts of fishing, recreation, tourism, shipping, coastal engineering, and materials reaching the sea as a consequence of activities on land. In many countries marine environment management strategy also includes establishment of protected areas to serve as sanctuaries for flora and fauna and to provide opportunities for scientific research and nonextractive recreation or tourism. In most countries there are institutions and mechanisms that can address each of the components of marine environment management but they are generally scattered among agencies at different levels of government and community administration. Often there is inadequate communication and sometimes there is mutual hostility and obstruction between such agencies.

The establishment of effective coordination is the critical first step in achieving marine environment management. All too often, rivalries between agencies within governments and between potentially complementary agencies of federal, provincial and local government are so great that coordination can only arise in response to a unifying threat. Most nations have to address deeply entrenched governmental and departmental traditions and power bases if they are to achieve a decision making structure that can cross the internal boundaries and stem the rivalry. Thus far this has been attempted by few nations.

In the international arena where issues can be addressed through the foreign affairs powers of sovereign or federal states, progress has been significant. The International Maritime Organization, the Food and Agricultural Organization, and the United Nations Environment Program's Regional Seas Program have developed conventions and protocols that, as they are adopted by more nations, are laying an international basis for management of marine environments of the high seas and of some impacts in the Exclusive Economic Zone and territorial seas of participating nations.

The challenge of conserving or restoring a sustainably healthy marine environment is to tackle the consequences of failure to consider limitations of the capacity of marine environments to absorb the impacts of the activities of large human populations on coastal lands. Solving the problems of poor practices of land and other natural resource management, urban and industrial waste disposal, and of poor agricultural use of soils and chemicals will involve a radically new approach to valuing and costing the use of natural waters and their resources.

The key lies in a collaborative approach between environment managers and economic managers that ensures that coastal and marine environments are used sustainably for purposes for which they are naturally suited.

Approaches to management of marine environments are evolving rapidly with growing awareness of the sensitivity of marine environments and of their actual and potential value in terms of culture, heritage, recreation, research, tourism, and mariculture. Recreation and the economic activities of providing for recreation through tourism have become an increasingly important economic motive for the management of marine environments providing a counterweight to the pressures of low cost waste disposal and unrestricted fishing.

Increasingly, marine resource and environment management is becoming a field of study and practice for a wide range of other professional disciplines. These include national park and reserve management,[2] fisheries and natural resource management[3] leisure, recreation and tourism, maritime law, regional planning, regional economics and geography. This book is intended to provide an introductory discursive text for such incoming marine and coastal area management professionals, graduate and senior undergraduate students. Most of the topics covered are the subject of substantial specific texts, a range of which is listed in the bibliography.

There are four main sections in this book. The first discusses general and theoretical issues that underlie management of marine environments and, where appropriate, contrasts them with terrestrial equivalents. It provides a brief history and seeks to define the basis upon which the management of marine environments and natural resources may be designed.

The second section concentrates on the practicalities of establishing and implementing marine management. It draws heavily on the UNESCO Coral Reef Management Handbook[4] but it presents some general guidance that is relevant to management of almost any marine environment.

The third section presents a case history of planning and management of the Great Barrier Reef Marine Park to illustrate the rapid change in perceptions of the relationship between human communities and marine environments.

The fourth section provides three examples, one of a collaborative treaty-based arrangement for addressing the complexities of the Mediterranean Sea and two of evolving approaches for the management of large marine areas arising from the island-based jurisdiction of less-developed nations. In the final chapter they are discussed, with the earlier example of the Great Barrier Reef, to identify some common themes.

The examples discussed in this book show a combination of two elements: measures at the scale of marine systems to address the process threat of pollution; and site-specific measures to protect ecological structure and to manage use and amenity. The detailed examples are taken from isolated areas recognized as having high natural environment heritage and research values and, consequently, high recreational and tourist values. These values have provided the motivation for communities and governments to address issues of marine environment management that apply far more widely.

The international programs discussed both started with a primary focus on pollution-related process conservation issues. Increasingly they are having to address issues in which the systemwide issues of pollution are heightened by site-specific values of specific areas for the maintenance of species that are fished or of biological communities that are fundamental to the amenity of leisure, recreation and tourism.

Chapter 1
Historical Background

From the earliest times, the sea has been a source of materials, a means of transport and a sink for wastes. Excavation of ancient middens shows that from the earliest prehistoric societies the sea has been a source of food and materials.[1] Legend and history record that transport and communication, made possible by mastery of the sea, extended the range of human communities and, through trade or conquest, gave access to new resources.[2]

Since times before recorded human history the seas have seemed vast and their capacities infinite in relation to human endeavor. The seas are the last frontier on earth. They have long been a source of myth, mystery, and great curiosity. The legends, prehistory, and archaeological fragments of many maritime nations tell of long voyages across the great oceans. Nordic warrior explorers and Celtic missionaries found landfalls across the Atlantic Ocean long before the American continent was "discovered." The master mariners of the Oceanian Islands traveled widely in the Pacific and colonized New Zealand. In the earliest history of exploration by Europeans, Florentine and Portuguese voyagers reached the Indian Ocean and the East Indies in the 14th and 15th centuries and returned with fragments of maps that indicated that Moluccan and Chinese navigators had explored most of the northern coast of Australia.[3]

Seafaring was a harsh calling for strong men. To venture to sea involved courage and determination and often heroic struggles in which the cruel sea was seen as a relentless mistress. Once the seafarer had set sail, communication was almost nonexistent. For those on land there was real uncertainty whether they would ever see their loved ones again and so they prayed "for those in peril on the sea."

The seas, virtually unexplored, were the last frontier, a great curiosity and a source of great potential wealth to the bold and ingenious. Mastery of the seas provided the basis of the fortunes of the seafaring nations of Europe. In addition to trade with distant nations, fortunes were built by taking fish, whales, seals, turtles, and seashells from seemingly inexhaustible, if increasingly distant, waters.

By the 15th century the seas were providing the means of long-distance transport of high value cargoes such as spices, silks and precious metals and stones from distant lands. By the middle of the 18th century, with the development of precise techniques for position fixing and navigation, and with improved design and construction providing larger and more reliable vessels, the hazards of sea travel were significantly reduced. The longer sea routes became the basis of the development and distribution of the large volumes of trade that flowed from the industrial revolution. The distant empires of the rapidly developing industrialized countries of Western Europe were important as sources of large volumes of raw materials and to a lesser extent as markets for resulting manufactured goods. By the close of the 18th century, military control of sea routes had become a matter of critical strategic importance on a global scale. The seas and their control became fundamental elements of the commercial and military policy of nations.

In the late 19th century the seas were still the romantic, dangerous, boundless last frontier.[4] The suggestion that humans were able to have significant widespread impacts upon marine environments was denounced as absurd.[5] The concept of needing to manage to avoid or minimize such impacts would have belonged in science fiction. Yet the science of the late nineteenth century provided the basis for the technological revolution of the twentieth century, which has increased scale and range of all human activities and brought major changes in human capacity to influence the seas. The relationship between humans and the marine environment has changed, shrinking the once boundless oceans and stressing their productive capacity.

Beyond the immediate reach of the technological, scientific, and legal world of the 19th century, there were human cultures on the coasts of all the continents and countless islands that had for centuries relied on the sea for dietary protein and other resources. In uncrowded temperate seas there were many navigational hazards but abundant opportunities were available in the seasonal fish stocks. Serious conflict between fishing activities was probably very rare. There was consequently little need for coordination. If management was needed at all, it was probably done formally or informally by the local community with little or no interest or participation from outsiders.

In ancient Roman law the concept was established of the sea, the seashore, and their products being common to all.[6] This concept has been passed on to most European legal systems and from them to legal systems in many countries that have used European models in developing statute law.

In estuaries and in tropical areas, such as coral reefs and mangroves, where resources and human populations may be highly concentrated, overuse or abuse could have serious consequences. Complex traditional management existed, developed presumably to conserve and share available resources and to avoid the damage and consequences of misuse. The relationship between humans and the marine environment was apparently simple and direct. If the marine environment deteriorated, the human food supply and resource base declined. Presumably the greater the dependency, the greater the consequences of damage. Most such regimes have been displaced by economic and social pressures resulting from the development of advanced technologies and the recent rapid growth of human populations. Some have been carefully documented[7] and shown to be based upon close observation and understanding of the biology of such areas. They contain a range of measures for managing human impact and sustaining the resource base. Traditional management regimes are generally based upon concepts of ownership and control by individuals or communities of areas of the coast, seabed or sea and the harvesting rights to resources in them.

MODERN CONCERNS

By the middle of the twentieth century, there was growing concern at the interlinked problems developing on land through the growth of human populations, the limited capacity of the global land mass to support continuing population growth, and the increasingly obvious impacts of human actions upon terrestrial ecosystems. The concept of the seas as the last frontier suggested opportunities for alleviating at least some of those problems.[8] Marine research and marine engineering were seen as the means of unlocking the resources of the oceans to provide large amounts of food and materials for human populations that threatened to exhaust the capacity of the 28% of the world's surface covered by land. Many nations and the international community placed special priority on marine research. Marine technologies resulting from this investment of research effort have fundamentally changed the traditional uses of the sea, navigation, and fishing.

In the late twentieth century, new technologies applied in ship and boatbuilding, marine propulsion, mapmaking, navigation, position fixing, communications, search and rescue, and weather prediction have dramatically reduced most of the perils and uncertainties of the mariner. Modern seafaring has so changed that it is now less hazardous than road travel. It is a calling for the technically competent who can remove most of the remaining hazards by careful attention to satellite and radio beacon position fixing, accurate charts, scheduled weather forecasts and equipment maintenance. Almost anywhere in the world, it is possible for the seafarer to transmit and receive instantaneous radio, television, or digital data.

New approaches or dramatic improvements to food preservation have made it possible to bring back quality fish from distant grounds. Physical and biological research has provided technology for finding fishing sites, and locating fish at those sites. These technologies have removed much of the commercial risk from fishing and have thus led to the development of high-capital commercial operations competing with local low-capital artisinal fisheries. Internationally, several nations have developed fleets of large long-range fishing vessels, which compete aggressively for the finite global fish stock.[9] In many cases their impact on available stocks has undercut pre-existing fisheries and eroded any traditional incentive for long-term sustainable management of particular areas owned and used by artisanal fisherman operating in a restricted range.

Individual fishermen, concerned at the decline of long-used fish stocks and areas during their fishing careers, may see the need for additional regulation of fishing, but natural caution tends to lead the majority to be slow to accept the need for management. A fishery usually starts and expands with little knowledge of the extent of the stock to be harvested or the consequence of harvesting. If the burden of proof for starting or expanding a fishery is minimal, the burden of proof demanded before restriction of fishing effort is accepted is usually so great that acceptable evidence takes the form of post hoc problems of biological and economic collapse. Cautious approaches to resource sharing and conservation, which limit catches to apparently sustainable levels that are less than the technical capacity of the fishing equipment are fiercely opposed.[10]

Changes in fisheries have coincided with changes in the impacts upon marine environments from land based activities. Increasing human populations generate increasing amounts of wastes. The structural changes of industry, agriculture and urbanization alter the patterns of drainage and generally increase the sediment load of freshwater running into the sea. In almost every field of terrestrial activity manufactured chemicals have become part of an increasing volume of wastes and by-products that enter rivers, estuaries, and coastal seas. Even the remote oceans have come under the influence of pollution. Tides and ocean currents result in pervasive mixing, which carries measurable amounts of man-made pollutants such as PCB's to deep ocean sediments and to animal tissues in the remotest areas of the oceans.[11]

The problems of fisheries and pollution have grown as new uses of the sea and new perspectives on management and resource sharing have emerged in an age of technology and leisure for the developed world. The original constituency of fishing and shipping interests has expanded. Recreation, tourism, research, education, and environment protection groups have developed with social and economic interests and concerns in the allocation and management of marine resources.

As the coastal plains of the world have become increasingly crowded, coastal seas have become prime recreation areas. They can offer naturalness and physical challenge that contrast with an increasingly built, modified, and controlled human environment. The seas can be areas for active or passive recreation, or for study and contemplation away from the pressures of urban life. Where people travel to take their recreation, beaches and shallow seas are prime sites for tourism. The application of new technologies to recreation and tourism has created new forms of use of marine environments. Aluminium and fiberglass dinghies, outboard motors, rotproof synthetic materials for sails, sheets and rigging, sailing boats from ocean-going yachts to sailboards, snorkel, SCUBA, and underwater camera equipment have created new, economically significant user groups, which have emerged to demand access to the resources previously shared by relatively few mariners and fishermen. They have greatly increased the numbers of people who become aware of and concerned by the deterioration of the marine environment.

GROWING AWARENESS AND ACTION

On the coasts of much of the economically developed and developing world, recreation and tourism have brought crowded waterways and competition for marine resources, for attractive day and overnight anchorages, for diving and fishing spots, for permanent mooring sites, for waterfront properties and for the business and employment opportunities offered by expanded use.

These changes have come to environments where legal and constitutional boundaries are complex, where, historically, many of the resources are regarded as common property and regulation has been piecemeal and often focused on resolution of single issues. Change has revealed the need for protection and conservation of marine resources and environments, the shortcomings of single issue management, and the consequent need for mechanisms to coordinate multiple, and often conflicting, demands upon marine resources.

Awareness of the need for protection and management of marine environments followed closely after public recognition that pollution and other forms of abuse were degrading and destroying terrestrial and freshwater environments. The 1950s and 1960s[12] made environmental degradation a matter of popular concern in the United States and Europe. At the same time, as scientific, particularly biological, research expanded, evidence accumulated on the deterioration of marine environments. This was often apparent in areas most accessible to human communities. Reports by Heyerdahl[13] of rafts of tar balls and other floating debris on the surface of the Atlantic Ocean far from land made marine pollution a major issue. This was compounded in the late 1960s by major oil pollution events as the supertankers *Torrey Canyon* and *Amoco Cadiz* were wrecked and discharged tens of thousands of tons of crude oil into the sea. Such events created an awareness of three needs: to control pollution of the sea, to conserve marine resources, and to protect some representative areas of marine environments.

An early, urgent focus of marine environment protection and conservation initiatives was on pollution. The priorities were clear, there was little dispute on the need to combat pollution, but a considerable amount of technical work was needed to develop standards and methods. The issues, being global, were suitable subjects for international conventions under the UN umbrella. International arrangements were established to address scientific technical matters such as standardising or cross-calibrating methods of determining pollution levels, for example, the Group of Experts in Sampling Marine Pollution (GESAMP).[14] The International Maritime Co-ordinating Organization (later the International Maritime Organization, IMO) concentrated on strategies for immediate response to catastrophic pollution and for longer term techniques to minimize pollution generally.[15]

Moves for the protective management of representative marine areas were somewhat slower to develop, possibly as a result of the legal difficulty of addressing issues that typically involve two or more levels of jurisdiction. A major impetus came with an International Conference on Marine Parks and Protected Areas convened in Tokyo in 1975 by the International Union for the Conservation of Nature and Natural Resources (IUCN).[16] Resolutions were passed at that conference that made a powerful case for a comprehensive, effective, and well-monitored global system of marine parks and reserves representing all coastal, longshore and oceanic ecosystems. Silva et al have reported on the listing of marine protected areas up to 1985.[17] They record that up to 1970, 118 marine protected areas were created, many of them adjuncts to terrestrial protected areas. 201 marine parks or reserves were created in the 1970's and 111 were created between 1980 and 1985 with a further 295 proposals listed as under consideration.

Around the world various legislative and administrative options have been followed in the establishment of conservation and protective regimes for marine areas. Silva et al list 91 categories of protected area used in marine environments.[18] The range reflects the difficulty of fitting marine area management into pre-existing legislative and management approaches. Within the variety, three basic legislative approaches can be identified. The first, exemplified by the Indonesian Marine National Parks system,[19] is to extend terrestrial national parks legislation to provide for the creation of national parks, reserves, or sanctuaries in intertidal and subtidal areas. The second, exemplified by the Malaysian Marine Protected area program,[20] is to broaden the scope of fisheries legislation to enable it to make specific provision for the protection of habitat of commercially significant species and for the conservation of marine resources generally. The third, exemplified by the Australian Great Barrier Reef Marine Park,[21] is to establish a coordinating agency to provide for conservation, sustainable development, and multiple use of the marine environment.

The National Park approach sees the establishment of Marine Protected Areas as a logical and conceptually minor extension of the terrestrial concept. Such areas, carefully defined, are set aside for the purpose of conservation of self-sufficient examples to represent specific ecological communities. They are generally managed by exclusion of activities that extract or harvest renewable resources and are seen as reserves for research and recreation that have minimal impact upon their environment.

The Fisheries approach sees the establishment of Marine Protected Areas as a logical and conceptually minor extension of the management of marine environments in order to sustain the harvest of renewable resources. Under this approach areas that are set aside are typically sites of immediate importance for spawning or larval development of commercially important species. They may not take much account of species that are not immediate targets for fisheries.

Coordination or multiple-use management offers a means to provide for new understanding, uses, or values of the marine environment. Introduction of multiple-use management is no easy panacea. It is most attractive to the new interest groups, recreational users, conservationists, divers and tourist operators who see it as a means to acquire influence in decisions concerning areas otherwise managed mainly for the immediate but long-established interests of commercial fishermen and coastal economic development. For established interests such power sharing may be less attractive since it generally brings loss of influence.

Marine conservation has become an increasingly important component of broader discussions of global resource conservation. The World Conservation Strategy[22] placed emphasis on the needs of marine environments. Increasing concern over the possible impacts of environmental abuse upon the role of the oceans in global climate and atmospheric gas balance has compounded concerns over the sustainability of ocean resource use. The Brundlandt Commission[23] recognized that the scale of the phenomena called for unprecedented measures to address management of marine environments on a global scale. General environment meetings such as the 4th World Wilderness Congress (WWC) and the 17th general assembly of the IUCN have adopted resolutions recognizing the urgency of the need for marine conservation generally and for the establishment of a representative system of marine protected areas.[24]

Chapter 2

Defining Management Areas

Effective management of marine environments, like management of atmospheric phenomena, involves cooperation to a degree that does not sit easily with scales of national decision making and concepts of independent sovereignty. Processes of terrestrial resource decision making, and the consequent machinery of policy, consultation, and management, developed on the basis of the concept of independent and staunchly defended sovereignty or jurisdiction. This concept has its origin in the apparent scales of terrestrial ecosystems and human capacity for defense of property and resources. It is geared to scales of tens of kilometers—a range over which the cause and effect and the costs and benefits can be perceived and discussed. The size of many older nations reflected these scales. Even in larger or federated nations, much critical, site specific decision making is done at a level of local government with a scale of tens of kilometers or less.

LEGAL BACKGROUND TO OWNERSHIP AND MANAGEMENT OF MARINE ENVIRONMENTS

For marine resource decisions, the machinery of policy, consultation, and management has only recently developed beyond the historical concepts of the limit of national jurisdiction being the 3-mile territorial sea. With the development of marine science and technology, particularly in relation to seabed mineral resources, there was a purpose for the definition of ownership. With knowledge of marine environmental systems and their vulnerability to pollution and over-exploitation, there was a purpose in environment protection. This has become apparent despite the fact that the scale, linkage, and dynamics of the ecosystem are so great as to obscure cause and effect or the costs that should be set against the apparent benefits of individual operations or impacts.

The 1970s and 1980s saw gradual acceptance of the need to make decisions regarding the use and management of marine resources. This was accompanied more slowly by the recognition that for many management issues the scale of decision making should be hundreds, if not thousands, of kilometers. To achieve this scale requires international mechanisms to identify or assign responsibility. The conventions developed at the UN Commission on the Law of the Sea[1] provided a framework of limited jurisdiction or sovereignty.

Development of policy for, and the implementation of, marine environment management often involves collaboration between nations that have different social and political systems and values and may indeed have a long history of territorial dispute and warfare. For a variety of reasons international treaties, or other less rigid arrangements for collaboration or support, are likely to be particularly important in achieving effective management.

The establishment of marine management regimes is often greatly complicated by issues of jurisdictional authority. Defining an area, identified on ecological criteria as requiring management, requires establishment of agreed landward and seaward boundaries. These boundaries will typically traverse the terrestrial and maritime jurisdiction of the coastal state concerned. In so doing they will involve local, national and, in a federal system, state or provincial levels of government. Such areas are likely to impinge on regions under the jurisdiction of a neighboring state and to extend far beyond traditional concepts of territorial waters into areas beyond national jurisdictions.

By definition, such an area usually consists, to a large extent, of wholly marine habitat in which the substrate, or seabed, is permanently covered by the sea. Frequently such areas also include intertidal habitats of a mainland or island coast in which the substrate is periodically covered by seawater. They may also include terrestrial components that extend above high water, such as islands or coastal wetlands important to the breeding of marine species.

The jurisdictional authority for a terrestrial area included within a marine environment management scheme is not usually a matter of dispute. However, the point of interface between terrestrial and marine realms is frequently a matter of jurisdictional sensitivity.

There are no simple rules for the establishment of boundaries for marine environment management areas. In each case the process usually involves considering a range of biophysical, geographic, and legal factors in order to develop a workable solution. This chapter discusses some principles underlying those issues. Social, economic, and political factors are usually also critical. These are considered later in a broader discussion of the nature of management problems in Chapter 4.

The concept of management of marine environments for purposes additional or alternative to fishing and navigation revealed areas of policy not clearly covered by existing law or legal conventions. Contentious issues such as seabed mineral potential and ocean pollution established the need to construct legal mechanisms to enable nations to define and agree on boundaries to the jurisdiction of each state over the seabed and waters adjacent to its coastline. These tasks were tackled at a series of UN Conferences on the Law of the Sea (UNCLOS), the first of which was held in 1958. Agreement on many of them was reached at UNCLOS 3 in 1982 with the signing of what is known as the Monetego Bay convention.

Some of the issues that were the subject of major debate at the UNCLOS meetings arise from the nature of sea/land interface. Some introduction to these issues may assist those approaching marine management from a background other than marine science, navigation, or fisheries.

The first critical issue is to establish a point at which the terrestrial jurisdiction of the coastal state gives way to its maritime jurisdiction. This is physically complex and is likely to be further complicated for federated states, such as Australia, Canada, or the United States, where there may be a need to resolve constitutional or jurisdictional issues between the provincial and federal levels of government.

The second critical issue is to establish a point at which the jurisdiction of the coastal state changes from that derived from the 3-mile territorial sea to that derived from more recent international conventions.

THE ROLE OF TIDES

The boundary between the land and the sea is dynamic. In most marine areas the major variable is that of the tides.[2]

Tides are a consequence of the gravitational pull exerted on the mass of the sea by the moon and the planets of the solar system. For the past century knowledge of the orbital behavior of the moon and planets has been applied, with adjustments to account for the effects of local topography, to produce reliable predictions of tidal variations for most of the ports of the world. The major influence comes from the earth's closest celestial body, the moon. In open waters the high point of gravitational attraction is at the point of the ocean surface closest to the moon. As the moon orbits the rotating earth, the high point reaches the submerged margin and the coast of the land mass of continents where it is observable as high tide.

The net gravitational attraction is greatest when the moon is aligned with the sun or other planets, and it is least when they are opposed. This accounts for regular and predictable variations in the time and amplitude of tides on daily, monthly, and longer cycles. The arrival of high tide at the continental margin may, however, be delayed some hours by physical obstructions to the progress of the tidal water mass, such as the British Isles in the case of the West European seaboard or the Great Barrier Reef in the case of northeast Australia.

The amplitude of the difference between low and high tide is also influenced by the topography of the coast. It is greatest where the tidal mass moves into an upwardly shelving and narrowing channel and least where the continental land mass faces the open ocean with little or no continental shelf.

In physical, geographic, and legal discussion, a number of terms are used to describe points in relation to tidal movements (Figure 2.1). Clearly, in normal circumstances on a coast, high water marks a point that separates areas that are regularly, frequently, and predictably covered by seawater from those that are not. Low water marks a point below which the geological substrate is generally covered by the sea. Depending upon the nature of the regulatory purpose, an interpretation of one or the other can offer a logical boundary. If the purpose is to minimize the extent to which land-based matters, such as agriculture or residential subdivision, will be complicated by issues arising from maritime legislation, the high water mark offers a logical boundary. If the purpose is to maximize the extent to which land-based matters, such as coastal reclamation or mining, may be covered by terrestrial legislation in the intertidal area, the low water mark will serve. There is thus a "gray area." Legal texts recognize that most functional boundaries between terrestrial and maritime jurisdictions lie within the tidelands, or the area between high and low water marks.

OTHER PHYSICAL INFLUENCES ON BOUNDARIES

Virtually all sea/land boundaries relate at some point to locations described in terms of tide. The imprecision of the description of locations described in terms of tidal levels can add an element of considerable uncertainty to the management of any area close to a tidally referenced boundary. The uncertainty and variability are compounded by other physical factors.

Tidal reference points form the most usual boundary definitions but the interpretation of these points at a location on the earth's surface can vary considerably as a consequence of weather and other interactive physical factors. Strong onshore or offshore winds in an area facing the open ocean will generally result in an observed high tide greater or less, respectively, than predicted. In each case there is a continuing process of normal events upon which there may be superimposed the impact of catastrophic or

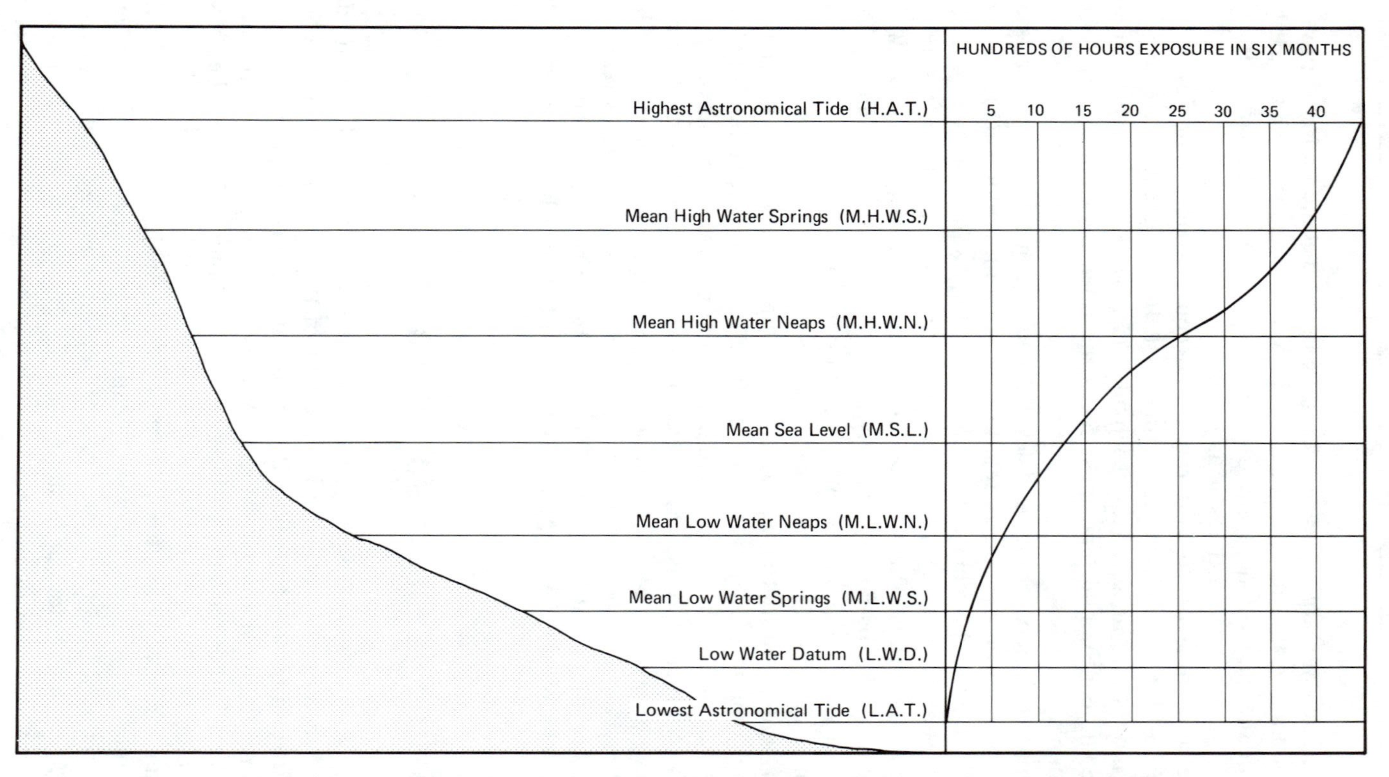

Figure 2.1 Diagram of some tidal reference points, indicating approximate hours of exposure of substrate at low water

episodic events that may cause extraordinary but substantial changes.

In areas such as the northeastern coast of Australia, where during the annual cycle the prevailing winds blow consistently from one direction for several months and then reverse, wind generated currents can move substantial volumes of sediment. Over a period of months, or even in the course of a single severe storm, this may have the effect of moving tidal reference points some tens or even hundreds of meters in relation to a fixed point such as a rocky headland or a coral reef (Figure 2.2).

A severe storm may cause sudden and dramatic changes to a coastline by removing or depositing sand masses. Depending upon the situation, such changes may be gradually reversed by the operation of normal processes over time.

River systems bring terrigenous sediments from land masses to the marine environment. As with wind-related effects, there are two scales of process. The first is the continuing process, exemplified by the Ganges River mouth in Bangladesh, where cumulative build-up of the sedimentary foreshores is stabilized by the growth of mangroves. Over a period of years the foreshore tidal reference points may move seawards by hundreds of meters. Severe storms may accelerate the process through major floods or temporarily reverse it through giant waves remodeling beach profiles.

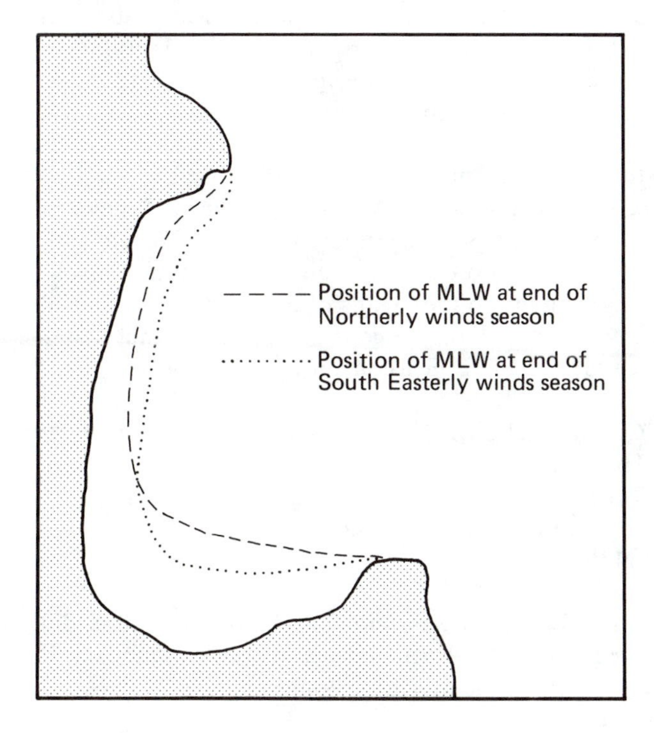

Figure 2.2 Diagram of a sandy bay indicating changes in the position of Mean Low Water as a result of prolonged wind action.

The situation is further complicated on a longer time scale by the processes of coastal geology.[3] The shoreline may move seaward or new islands may be created by lava flows or by tectonic lifting of a continental margin. Examples may be seen in the Galapagos Islands, Hawaii, Iceland, and the New Britain coast of Papua New Guinea. The shoreline may move landward through the process of declination, or downward tectonic tilting of a continental margin as is the case in the eastern seaboard of North America. On the coast of Bangladesh there is a precarious balance between the accumulation of land through sedimentary processes and the loss of land through declination of the coast.

BACKGROUND TO THE CONVENTIONS DEVELOPED BY UNCLOS

Under customary international law, derived from the practicalities of the eighteenth century, which took account of the effective range of fire of a contemporary cannon, a distance of 3 nautical miles was generally used to define the extent of the territorial waters of coastal states. It was accepted internationally that a coastal state had the right to exercise exclusive control within its territorial sea. By the 1950s a number of coastal states had taken legislative or executive action to assert rights of sover-eignty or jurisdiction over their adjacent continental shelves to a distance of 12 miles. Some nations, particularly in South America, claimed greater distances, up to 200 miles.

The issue of the extent of the territorial sea was addressed at the first UN Conference on the Law of the Sea (UNCLOS 1) held in Geneva in 1958, although general agreement on the matter was not achieved until UNCLOS 3 in 1982. UNCLOS 1 resulted in the adoption of four conventions that provided for agreement on the means of defining the territorial sea of a coastal state and the rights of the coastal state and other states within the territorial sea. The UNCLOS 1 conventions are:

Convention on the Continental Shelf
Convention on the High Seas
Convention on the Territorial Sea and the Contiguous Zone
Convention on Fishing and the Conservation of Living
Resources of the High Seas

The Convention on the Territorial Sea and the Contiguous Zone did not establish an agreed extent of the territorial sea of coastal states but it did establish the manner in which the maritime zones adjacent to a coastline were to be measured.

Lumb[4] developed brief definitions of a number of maritime concepts that arose in UNCLOS discussions. They are presented here with some modified to reflect agreements incorporated in the Montego Bay convention and accepted as representing customary international law following UNCLOS 3 in 1982:

TIDELANDS: The area of land covered by the ebb and flow of the tide, that is the area that lies between high water mark and low water mark.

INTERNAL WATERS: Waters that are either entirely enclosed by land (such as lakes) or that, bordering on the open sea (such as bays), are enclosed within a baseline drawn in accordance with a series of rules summarized in the next definition.

BASELINE OF TERRITORIAL WATERS: This follows the low water line along the sinuosities of the coast of the mainland and islands of the coastal state. Where there are arms of the sea (estuaries or harbors) or significant indentations of the coastline (such as gulfs or bays), a practice applies to determine the conditions under which the baseline is drawn across the mouth to join the low water line on either side. The convention, known in Latin as "intra fauces terrae" (within the jaws of the land) applies where the depth, or distance from the mouth of the indentation to the low water mark, is significantly greater than the length of the opening to the open sea. Under this practice the area of such a bay must be at least as large as a semicircle whose diameter is a line drawn across the mouth of the bay. Where the mainland coast is fringed with islands in such a way as to assimilate the waters lying between the coastline and the "fringe" to the regime of internal waters, the baseline may be drawn as a series of straight lines connecting the low water lines of the fringing islands. An island is an area of land, above water at high tide, which is surrounded by water. The low water line of an island constitutes a baseline. In contrast, a low water elevation is an area of land surrounded by water, which is above water at low tide but not above water at high tide. The low water line of a low water elevation does not generally constitute a baseline. It may do so where it is wholly or partially situated at a distance not exceeding the breadth of the territorial sea from the mainland or an island.

TERRITORIAL WATERS: The area of water that extends from the baseline (accepted generally as the low water line) to a distance from the shore was commonly fixed at 3 nautical miles. UNCLOS 3 in 1982 gave to coastal states the right to extend the breadth of their territorial sea to 12 nautical miles. Under UNCLOS 3, coastal states undertook to prepare charts identifying the baseline for their territorial waters and thus the limit of internal waters. States cannot prohibit the innocent passage of foreign ships through their territorial waters, although they can regulate it.

CONTIGUOUS ZONE: This is an area extending no more than an additonal 12-nautical-miles from the outer boundary of the 12 territorial waters of a coastal state. Within a contiguous zone a coastal state may exercise a jurisdiction over foreign vessels in relation to customs, fiscal, immigration or sanitary matters (Convention on the Territorial Seas and Contiguous Zone, Article 24).

HIGH SEAS (or International Waters): These are waters of the world's oceans and seas that lie outside territorial waters. They are open to use by all nations and cannot be appropriated by any one nation, although they are subject to certain rules so far as use is concerned.

THE CONTINENTAL SHELF: The seabed and subsoil of the submarine areas adjacent to the coast of a mainland or an island, extending to the edge of the continental margin or to the 200 mile limit.

EXCLUSIVE ECONOMIC ZONE (EEZ): A zone extending to a distance of 200 nautical miles from the baseline, or to the edge of the continental shelf if this is more than 200 nautical miles from the baseline. Under the terms of the treaty negotiated at UNCLOS 3, coastal states have jurisdiction over the EEZ with regard to preserving the marine environment and managing living and non living resources.

EXCLUSIVE FISHING ZONE (EFZ): A coastal state which does not proceed to full implementation of an EEZ may declare an EFZ within areas covered by its entitlement to an EEZ. By EEZ or EFZ, the coastal state asserts its right to harvest the maximum sustainable yield of the living natural resources of the area and its responsibility for managing them. If the coastal state is not in a position to exploit the available biological resources, it is obliged, for a reasonable financial return, to make them available through license or joint-venture arrangements to grant access to the fishing fleets of other states.

CONTINENTAL SLOPE: The zone commencing at the 200-meter depth contour (that is the limit of the "physical" continental shelf where there is usually a marked increase in "fall away") to the edge of the continental rise (which varies between 1200 and 3500 meters in depth).

DEEP SEABED OR ABYSSAL PLAIN: This is the area of the seabed beyond the continental margin. The seabed, ocean floor, and subsoil beyond the limits of national jurisdiction are however a distinct zone where, under the terms of UNCLOS 3, the exploitation of mineral resources is to be managed by a new body, the International Sea-Bed Authority. Nevertheless some nations, including the United States, have not accepted the deep seabed regime and there is thus doubt that it represents customary international law.

The legal and technical complexities of applying the conventions to draw baselines are complex. Alexander, Prescott, and Beazley[5] have provided a comprehensive discussion. Some of the issues may be clarified by Figures 2.3 which illustrates the territorial waters of an imaginary coastal state and the definition of a baseline for the purposes of external powers in the case of a land mass with coastal islands. Figure 2.4 illustrates some of the conventions which have been applied to define the extent of the baseline of jurisdiction of a coastal state.

Figure 2.5 illustrates the application of the 3-mile Territorial Sea, the 12 mile territorial sea, the contiguous zone and the boundary of the EEZ

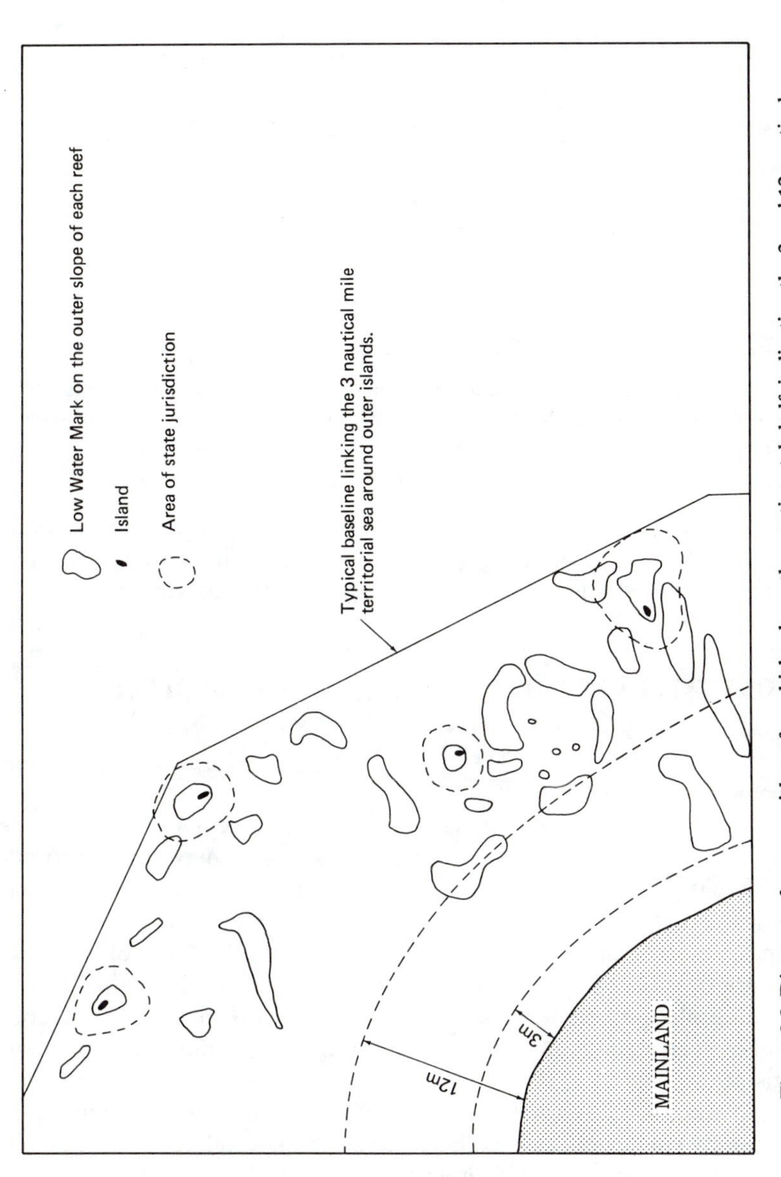

Figure 2.3 Diagram of a coast with reefs and islands on the continental shelf, indicating the 3 and 12 nautical mile territorial sea drawn from the mainland, the 3 nautical mile territorial sea drawn from the islands and the archipelagic baseline linking the 3 nautical mile territorial sea around the islands.

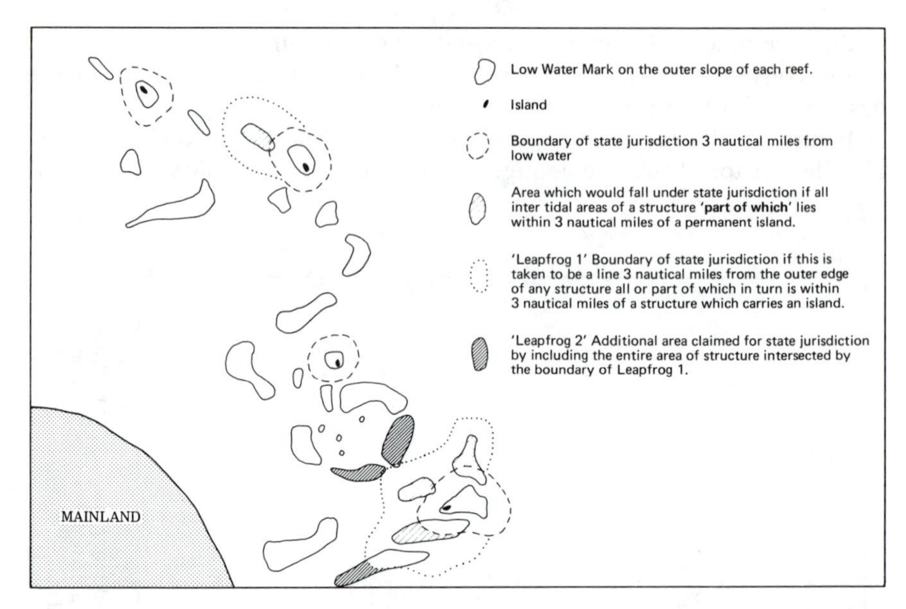

Figure 2.4 Diagram of a coast illustrating the "Leapfrog 1" and "Leapfrog 2" concepts whereby the baseline of the territorial sea is taken to extend to further inter tidal structures.

defined by a line A: 200 nautical miles from the continental low water mark, B: 200 nautical mile from a baseline incorporating an island within the contiguous zone; and C: along the margins of the continental shelf.

THE INTERNATIONAL LEGAL BACKGROUND TO ESTABLISHING MARINE ENVIRONMENT MANAGEMENT

Protected areas may not be established on the high seas unless all states agree by treaty. The International Sea-Bed Authority may be able to apply some measures for the protection of some areas of seabed beyond national jurisdiction.

Marine protected areas may be established within the EEZ of a coastal state by application of appropriate maritime legislation. Measures to extend or integrate management of such areas landward to include adjacent intertidal or terrestrial areas will generally require specific or complementary legislation.

Generally, marine environment management initiatives are likely to be undertaken by single states or, collaboratively, by several states using powers derived from their jurisdiction within their EEZs.

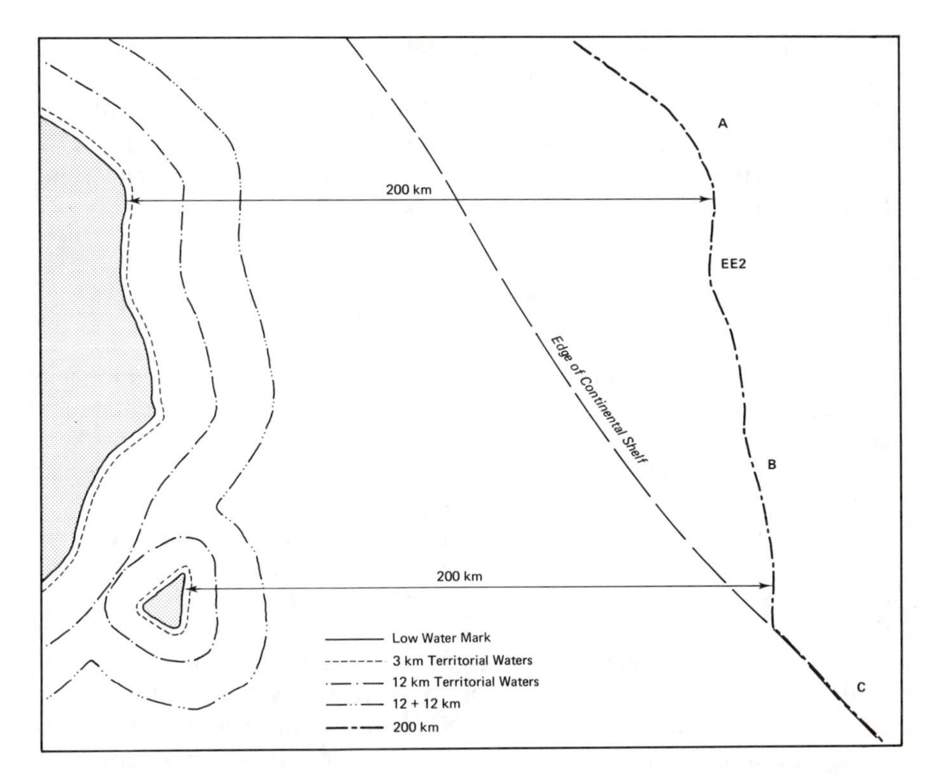

Figure 2.5 Diagram illustrating various forms of territorial claim based upon distances from low water and a 200 nautical mile EEZ derived in part from low water and in part from the edge of the continental shelf

The UN Environment Program (UNEP) established the Regional Seas Program to provide a regional basis for addressing concerns about the environmental problems of the oceans.[6] As a UNEP initiative the Regional Seas Program provides a coordinating mechanism for nations that may have substantial political differences to collaborate in seeking solutions to problems of inevitable mutual consequence.

Within the Regional Seas Program the coastal states of geographic regions have adopted action plans after considering information on the nature and extent of the various threats to the marine environment of their region. An action plan is generally followed by the adoption of a convention which specifies general arrangements for such matters as pollution control, monitoring, fisheries and other aspects of environment protection and restoration. The initial focus of Regional Seas initiatives was primarily concerned with the impacts of pollution but the regional action plans have become increasingly involved in management of related ecosystems such as coastal wetlands and catchments.

The Regional Seas regions are illustrated in Figure 2.6. The first to be established, in 1975, was the Mediterranean. Part of the action plan for the Mediterranean has included establishment of a system of protected areas that are recognised by the first international legal instrument directly devoted to marine protected areas, the Protocol Concerning Mediterranean Specially Protected Areas (UNEP, 1982).

CONCLUSION

The marine areas most directly exposed and vulnerable to human impacts are generally those of coastal and shallow water areas of continental margins. The creation of marine environment management systems for such areas often raises complex and contentious issues of jurisdictional rights and responsibilities.

For many marine ecosystems, the functional ecological boundaries encompass areas under two or more human jurisdictions. This is clearly the case where an ecosystem extends into waters of areas where the jurisdictions of several coastal states abut. It can also be the case within coastal states that have a federal system of government. The federal government is the competent body in relation to matters involving international treaties, agreements, and legislation. Nevertheless, in such states it is commonly the case that internal regulation (covering activities in internal waters and the area covered by the old 3-mile territorial sea) lies wholly or partially under the jurisdiction of the provincial government, whereas external regulation and defense of the 3-mile territorial sea, and regulation of all activities in the remainder of the territorial waters and the EEZ, lie under the jurisdiction of the federal government.

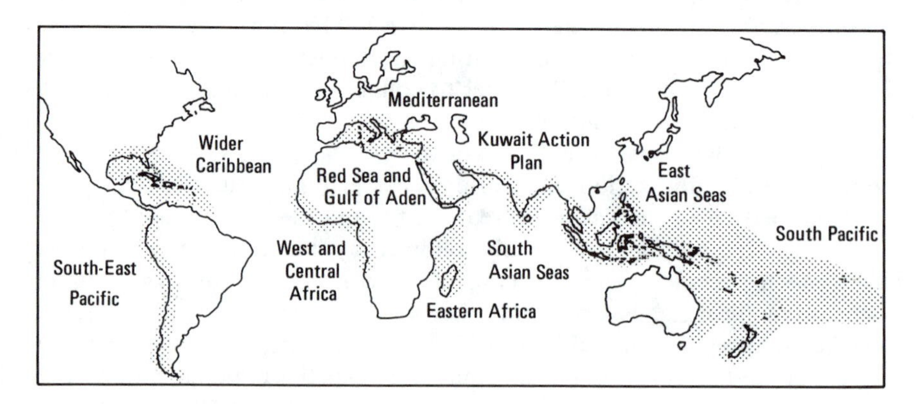

Figure 2.6 Diagram showing the areas under the UNEP Regional Seas Programme

Understanding and achievement of agreement over issues of jurisdiction are essential for the creation of viable and effective marine environment management regimes. UNEP Regional Seas action plans offer a framework for regional collaboration, particularly in situations where neighboring nations have profound current or historic differences.

Chapter 3

Nature Of Marine Systems Relevant To Management

The conceptual framework and traditions with which we approach the management of human activity have developed over millenia of experience on the land and coastal fringes of the sea. They have developed in environments where almost all life processes occur within 100 meters of the surface of the earth. Terrestrially derived concepts of space, seperate entities, natural boundaries and the dynamics of natural systems can be inappropriate or irrelevant in the sea.

THE SCALE OF MARINE ECOSYSTEMS

On land, we are used to the concept of biological communities associated with sites bounded by identifiable and often distinctive geological features such as rivers, mountains, the sea, the extent of a catchment, or the occurrence of a particular soil type. Generally, such biological communities can be considered as two dimensional and mostly site-attached; a layer of soil with interacting animals and plants on or within a few meters of the land surface. In the third dimension, the air column above such sites is a neutral transport medium enabling passive distribution or transport of pollen, spores, seeds, and drifting or flying animals but providing no nourishment or sustenance. Plants and animals cannot grow, feed and reproduce without using resources formed on the surface or generally, without direct contact with the surface.

Precise, permanent geographic boundaries that contain or separate biological communities and processes are rare in marine environments. The third dimension, the water column above the seabed, is active. It nourishes and sustains communities of plants and animals. Some drift or swim perpetually others are the planktonic spores, eggs and young of most of the species whose adult forms settle on the substrate. Above the essentially two-dimensional community of the seabed is a three-dimensional water column containing its own communities and the elements of many seabed communities. The water mass above an area of sheltered sandy seabed is likely to carry the genetic material, spores, and larvae for a range of communities whether rocky, sandy or mud substrate, exposed or sheltered, shallow or deep.[1]

The water mass is rarely static. It moves with the wind, with the tides and with residual currents. As it moves its biological communities change with the biological processes of photosynthesis and respiration, the feeding, excretion, and death of grazers and predators, the birth and development of larvae from newly discharged eggs and sperm, and the settlement of mature larvae. The chemical nature of the water mass alters continually. It may gain nutrients from ocean/atmosphere interactions, runoff from land masses, deep ocean upwellings, and bacterial breakdown of detritus. It may lose them as they are absorbed by plants and animals or as they are adsorbed onto the surface of particles that settle onto benthic sediments.

The effects of the movement and variation of major currents can be spectacular. If a warm current intrudes into an area usually occupied by a cool water mass, it will bring subtropical and even tropical forms.[2] Larvae may settle and mature to adulthood, displacing cooler water forms. Fish fauna may change dramatically, as is the case with the sardine/anchovy shift in Peruvian waters.[3] These water current changes and reversals may be catastrophic to the established fauna and flora, but they represent opportunities for intruding species.

The biological communities expressed in the water column on and above a point in the seabed may be regarded as dynamic and four dimensional. At any point of time they represent the horizontal and vertical integration of upstream events in the water mass over a variety of scales of space and time, most of them vastly different from any comparable situation on land.

CONNECTIVITY

The linkages of the active water column lead to fundamental differences between terrestrial and marine environments. We are used to the daily and seasonal processes of nutrition and growth being largely site-related on land. The energy of sunlight is fixed by plants, which are directly or indirectly attached to the substrate. Most animals move in search of food.

Most plant nutrients are obtained from the soil locally or from materials borne from a limited upstream catchment. Energy and nutrients, thus fixed, may be moved considerable distances by displacement of seeds, insects, and birds blown by the wind. They may also be borne away and dispersed by a flood, or released to the atmosphere as smoke from a fire. There are major migrations of birds and large mammals that transfer energy stored in a rich feeding season in high latitudes or seasonal grazing lands.[4] Nevertheless a major part of the biological products resulting from photosynthesis remain at or fairly close to the point of production.

To a considerable extent, the biological processes occurring within a geographic unit on land are self-contained. Unless adjacent sites are related in a drainage catchment, or on a significant migration route, they are not likely to be closely interlinked. The transfer of materials between sites in different catchments was rarely a substantial issue in terrestrial environment management until the phenomenon of acid rain became apparent.

In the sea most photosynthetic fixing of the energy of sunlight is carried out by the drifting plant cells of the phytoplankton. These plants and many of the animals are effectively part of the water mass. They move with it and within it. The communities that are attached to the substrate depend upon the moving water mass to bring nutrients and food to them. Massive transfer of materials between sites is a basic process of marine community dynamics. Isolated sites are the exception rather than the rule. Over a large area affected by a particular water mass, the same or very similar benthic communities are likely to occur wherever appropriate conditions of geological substrate, water depth, and shelter coincide. Such locations may be many miles apart, but the communities they support are likely to be genetically very closely linked.

SEAWATER AS AN ENVIRONMENT FOR LIVING TISSUE

The sea provides benign conditions for biological cells. Sea water is chemically buffered, it provides nutrients and absorbs metabolic by-products. Changes in temperature occur gradually. In open waters at a depth of 1 meter, a change of 1 degree Celsius in 24 hours would be regarded as substantial. Most harmful solar radiation is absorbed in the first metre of the water column.[5]

In air, living systems must be able to cope with extremes. They must maintain a moist, chemically buffered environment for the cells that make up their tissues. They must have the means of removing or neutralizing toxic metabolic wastes and of coping with large fluctuations of temperature and solar radiation. Wind may cause dessication and freshwater from rain may dilute the cellular environment.

DISTINCTION BETWEEN INTERTIDAL AND SUBTIDAL

The most familiar marine environments are those of the margins of land that are atypical in that they face many of the problems of terrestrial existence. In many parts of the world there are rich intertidal communities of attached plants and animals and of dependent free-living animals. Familiar as they may be to coastal humans, intertidal communities contain exceptional marine species that are able to cope with problems that do not arise for fully marine species. Their environment changes as the tide rises and falls. When the tide is in, they benefit from a rich environment with high levels of nutrients and light in which plants flourish. Food chains of animals are supported by the attached plants and by detritus and organic matter from land. When the tide is out, these communities share with terrestrial plants and animals the need to solve the problems of dessication, temperature shock, disposal of metabolic wastes, exposure to solar radiation, and occasional inundation by freshwater, often with accompanying loads of silt, from rain storms or coastal runoff.[6]

Shallow subtidal communities can also be superficially and deceptively similar to the terrestrial environment. High light levels and stable areas of the seabed support attached photosynthetic communities; forests of algae, meadows of seagrass or coral reefs. These in turn support grazing animals. They function in a manner similar to the terrestrial norm of moving animals supported directly or through predation by fixed plant communities. They also contain species that function according to the norm of the fixed or free-living forms of the deep seabed and water column which take their food from the plankton and suspended detritus of the water column. Despite similarities to those of terrestrial environments, most species in the biological communities of intertidal and shallow subtidal generally have reproductive strategies similar to those of other marine environments.

REPRODUCTIVE STRATEGIES IN THE SEA

For most species in the relatively benign environment of seawater, parental investment in an individual larva extends no further than a small amount of metabolic energy for the larva to develop to a point at which it can feed itself within the planktonic community. Many species appear to have behavioral or physiological adaptations to coordinate spawning to increase the chance of eggs being fertilized. Beyond this, the reproductive strategy is to maximize the numbers of eggs and sperm discharged into the plankton. Many, often millions, of eggs are produced by females of species that have planktonic larvae. The resulting larvae drift in the water column taking their chance of being fertilized by a coincident sperm and drawing their metabolic and developmental needs from materials dissolved in or supported by the surrounding water. As they drift and develop the young are transported by the water mass. A minute proportion survives to be carried to an area suitable for adult life.

In contrast, most terrestrial plants and animals devote substantial energy to producing relatively few large seeds or yolky eggs which can function as energy reserves and survival capsules until the young reach a stage at which they have a reasonable chance of surviving. Many species have elaborate physiological and behavioral mechanisms, such as dormancy, viviparity or parental brooding and caring behavior to protect their young and to ensure that the next generation has a reasonable chance of finding a habitat or environment suitable for adult life.[7]

The problems of larval survival are compounded for terrestrial species where adult habitats are geographically limited and functionally separated from similar areas. The geographic isolation of such parent terrestrial communities can pose problems for genetic mixing with other communities of the same species. The harsh, isolated and fragmented environments of such species thus tend to favor adaptation, specialization and reproductive isolation. Specialized terrestrial species are relatively common but they are vulnerable to extinction if their specialized habitat is destroyed or significantly modified by human activity. The endangered species endemic to a specialized habitat is a fact that is fundamental to much of terrestrial conservation philosophy. Conservation of rare and endangered terrestrial species is an obvious priority that can often be immediately addressed in relative isolation from the mainstream of human activity management by excluding incompatible human use from relatively small but clearly definable areas.

A few marine species, particularly some molluscs and fish, reproduce by caring for relatively small numbers of eggs or young rather than by planktonic larvae. Some of these may be endemic to specialized habitats. Marine-dependent species such as sea birds and secondarily marine air-breathing species, the marine mammals and marine reptiles, care for their young and are particularly vulnerable when breeding on land or in known areas of sheltered and generally shallow sea. Such species present the same management issues and challenges as endangered species on land.

In contrast marine species with planktonic or free swimming larvae are rarely site-dependent. The demonstrably endangered endemic species and its threatened habitat are rarely issues in marine conservation. Admittedly, the logistic difficulties of most marine studies have the effect that less effort has been put into species inventory in marine environments than on land, but it is remarkable that there are no records of apparent twentieth century global extinction of a marine fish or invertebrate.

The richness, relative permanence or regular seasonal recurrence of intertidal and subtidal communities at specific locations means that, where accessible, they are often prime sites for fishing and recreation. Their use by humans can be controlled in the same way as for terrestrial sites, on the basis of regulations applying to areas that may be precisely defined in relation to landmarks or specific points on the surface of the earth.

The combination of long linkages and the problem of large-scale studies in the marine environment means that it will rarely be possible to demonstrate that management control of a specific limited intertidal or subtidal area makes, or is likely to make, a specific contribution to the survival of any fish or invertebrate species. Such control can protect the human amenity values, such as recreational opportunities; reduce conflict between forms of use; set aside areas for contemplation and research free from the influences of fishing or collecting; and protect sensitive sites, such as known nursery areas for juvenile fish and invertebrates.

LARGE GEOGRAPHIC SCALES

The movement of water masses and their associated chemical and planktonic characteristics may extend over vast distances. A planktonic larva carried for 28 days at a net speed of 1 knot will travel almost 600 nautical miles. A spawning or pollution event may thus have significant effects many miles from its point of occurrence.

In terrestrial environments, most ecological communities can be, addressed by survey and management scales of 10^1 to 10^4 meters. These scales are generally appropriate for fixed or territorial components of intertidal and benthic communities but for planktonic species scales, of 10^4 to 10^6 meters are appropriate while for nektonic and migratory species scales of 10^4 to 10^7 meters apply. The range of life cycle strategies can dictate a range of scales. For many marine animals and plants a fixed phase produces gametes or propagules that produce a larva that develops while it is transported in the plankton by ocean currents. Some, such as jellyfish, operate in the reverse with the planktonic sexual stage budded or discharged from a fixed asexual stage. A few animals and plants do not have a planktonic or pelagic phase and some pelagic or migratory species have specific breeding sites. Many species have no fixed or site-dependent stage in their life cycle. A benthic species that broods its young may be addressed by a small area. At the other extreme, a whale that makes annual migrations between Arctic and Antarctic must be addressed at a global scale. Figure 3.1 illustrates the geographic scales of plant and animal distribution in the sea in relation to the attached or territorial and the distribution phases of life strategies. It can thus illustrate the apparent scope of site-specific management strategies.

Type A. A fixed or territorial adult phase with a strategy, such as brooding, viviparity, or attached yolky eggs which does not involve planktonic larvae. Such species may have a limited adult and larval range. Site-specific area protection may be critical for species management in a strategy similar to the terrestrial national park.

Type B. A fixed or site-dependent phase with planktonic larvae or a large migratory range. The site may be a distinctive benthic structure or community, such as a coral reef, or it may be the nesting site for turtles

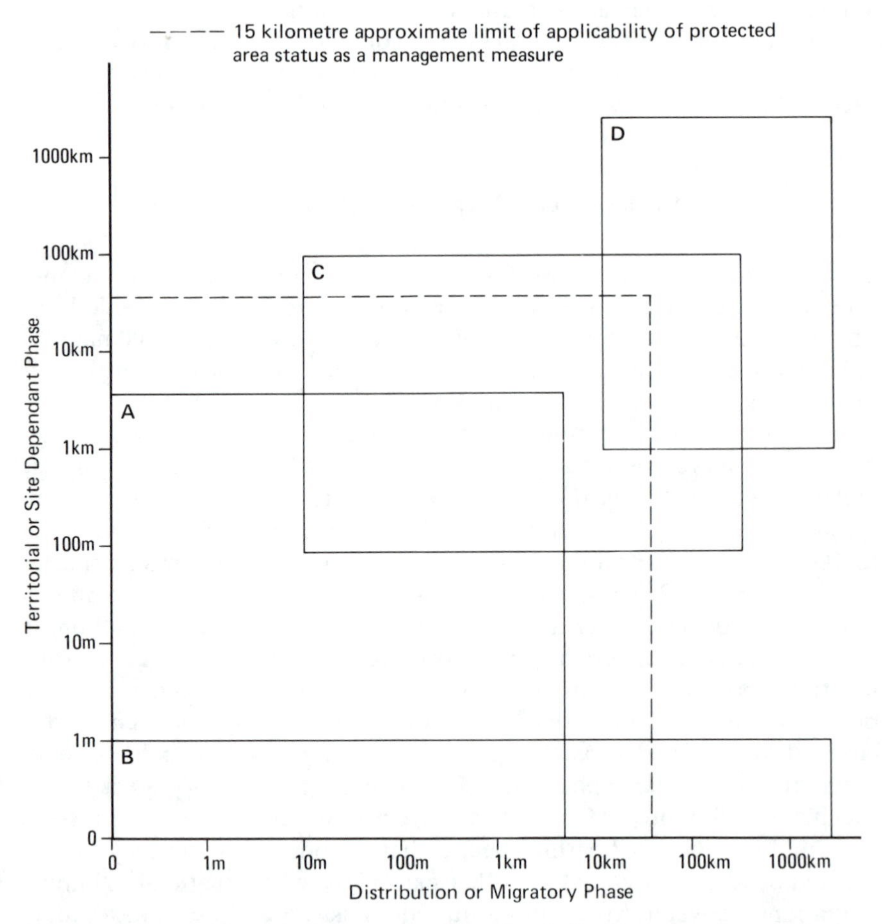

Figure 3.1 Diagram illustrating four marine life cycle relationships relevant to marine area managment. A: Fixed or very restricted movement in the adult phase with no planktonic or pelagic larval or juvenile phase. B: One phase fixed, the other planktonic or pelagic. C: Limited territorial range of adult, extensive planktonic larval range. D: All phases pelagic or demersal.

and birds, the calving area for whales or spawning area for pelagic fish. In such a situation site-specific protection contributes to species management and may address a critical life-cycle stage but it will generally need to be supported by more general measures over the range of the species concerned.

Type C. A limited adult territory and a planktonic larval phase—e.g. shore crabs, lobsters and territorial reef fish—or a nursery area (generally inshore) for planktonic or benthic larvae of species with a large adult range. Site-specific management is an effective approach where suitable areas for adult territory or larval nursery are limited, as may be the case for species that use estuaries, mangroves, coral reefs or salt marshes. Otherwise, species management is largely a matter of conserving processes and environmental quality throughout a significant proportion of the distributional range.

Type D. Pelagic or planktonic adult with planktonic larvae. Site specific management is unlikely to make a substantial contribution. Conservation and management are process related.

PROBLEMS OF OBSERVING THE MARINE ENVIRONMENT

The problem of the large scale of marine systems is compounded by the fact that field investigation in marine environments is seriously restricted by the aquatic medium. On land, a single researcher can spend days living in the study environment achieving hours of direct onsite study by day or night. Through continued presence in relative safety and comfort, such a researcher has the opportunity to observe and record rare events and to appreciate the context of the interactions of the components of the studied environment. On the surface and unaccompanied, the observer can travel and observe on foot or by vehicle and can, unless obscured by dense vegetation or cloud, see objects hundreds of feet away and study them through binoculars or long-range camera lens.

Above the surface of the land, the observer can identify major vegetation types or count large animals from an aircraft. A wide spectrum of light wave-lengths can be reflected from the land surface and recorded by aerial or satellite remote sensing. Remotely sensed data can be analyzed to deduce a considerable amount of information on the extent and dynamics of catchment areas, surface slope, texture, soil type, biological communities and human impacts.

SCUBA diving has enabled humans to enter shallow marine environments and to observe a small part of them directly. But such are the limitations of the technology and human physiology that the diver is a transient and very temporary observer. A diver may be able to conduct *in situ* experiments with planktonic forms, but generally the planktonic plants and animals are of microscopic size not amenable to direct observation in the field. The diver is effectively restricted to limited observations of the essentially two-dimensional community of the shallow seabed. In a day, it is unreasonable to expect an observer using SCUBA equipment to make detailed studies for more than two periods of an hour or so. During such studies the diver should be accompanied by a buddy and should have at least one person on surface standby supervision. Thus at best, three people are required for one person to achieve a total of four person hours of direct on site study in a day. This figure is usually further reduced when it is necessary to account for the person hours of crew for a mother ship required to support diving operations at a study site more than a few miles from land. Observations can be made at night, but they often require more surface support than daytime operations. In any case, depending on the depth, prudent diving practice would still restrict the diver to a maximum of about 4 hours in a 24-hour period.

The aquatic medium greatly limits the capacity to make long-range or remote observations. In the clearest of water the maximum range of vision for a diver or a camera is about 60 meters. In many inshore waters the range of vision is usually less than 7 meters and rarely reaches 15 meters. Swimming energetically, or towed by some form of vessel, a diver, using SCUBA or snorkel equipment, can in an hour carry out a reconnaissance or superficial survey over a track of 2 to 5 kilometers by 5 to 10 meters.[8]

The potential of aerial photography and satellite remote sensing are severely limited in comparison to terrestrial environments because the water column absorbs light, particularly in the shorter infrared to green wavelengths.[9] In clear waters there can be useful reflectance from the seabed beneath shallow water but this declines rapidly, with little reflected signal in depths greater than 15 meters. If the water column is turbid, as is frequently the case in coastal waters, the effective depth of vision is further reduced.

Considerable progress has been made interpreting the limited data available within the capability of satellite remote sensing. It has been possible to prepare maps and to interpret physical conditions and probable associated biological communities of shallow seabed areas of the Great Barrier Reef Region and other North Australian waters. Despite this, the marine researcher and manager is generally faced with very large natural systems but is armed with a very limited capacity for reconnaissance and interpretation with which to target very limited capacities for detailed field investigation.

DIFFERENT MANAGEMENT SCALES

Any human lifestyle on land, with the possible exception of low level hunting and gathering, involves deliberate alienation and fundamental alteration of areas of originally pristine environment. This is done for purposes ranging from pioneer agriculture to urban and industrial development. The relative isolation or lack of interaction between sites has made it possible to deal with issues individually. Typically, human activities are regulated on the basis that a range of uses can be accommodated by subdividing an area or a resource and allocating subunits for different purposes with, generally, little interaction between the subdivided sites. Problems or disputes generally occur on the scale of kilometers or tens of kilometers so mechanisms to manage these can generally operate at the level of the local community or local government. They may become national government issues if there is profound local disagreement or if a plan or actions at the local level appear to have significant broader political, economic, or ecological implicatlons .

In the sea, the majority of uses are still at the level of hunting and gathering. Alienation and fundamental alteration of the environment of areas of the seabed or volumes of the water column have not been substantial issues until very recently. Now, mariculture, seabed mining, reclamation, tourism and some forms of trawl or dredge fishing can all alienate or alter areas of seabed. Pollution can modify the water column. Some aspects of activities may be regulated by subdividing areas and resources and allocating subunits for different purposes. Nevertheless, such are the linkages in marine systems that there may be substantial interactions over long distances. Problems or disputes are likely to occur on the scale of tens to hundreds of kilometers and to require resolution at the national or international level.

ALIENATION AND PERCEPTUAL HURDLES

A major difference arises from the fact that when humans pass through the air-water interface they are in an alien and mysterious environment. On land, it is easy to demonstrate the life cyle of many of the animals and plants. However, in the sea the larvae of most species are microscopic and planktonic. The settled young are usually very well hidden and often have a form and habitat very different from their adult forms. The dynamics of populations of marine animals are complex for the marine ecologist. To the nonspecialist, marine animals such as lobsters, barnacles, sea urchins, and fish seem to materialize from nowhere and their numbers seem to fluctuate for no apparent reason.

The strangeness of marine animals is increased by centuries of anthropomorphic appreciation of the form and life histories of land animals. Folklore, song, and particularly stories and books for children, lead people to observe and admire land animals and plants for their beauty of form, their adaptations and their behavior. Mammals and birds, in particular, inspire devotion, empathy, and strong popular support in the face of actions that might threaten them. In the sea, only whales, other marine mammals, turtles and seabirds inspire similar popular sentiment. They breathe air and care for their young. They operate on a basis that can be related to that of familiar land animals, although none of them is typical of the bulk of marine life.

Fish, worms, seastars, jellyfish, or scallops do not so easily evoke public empathy. They are cold-blooded, they can breathe in water which drowns mammals, most do not care for their young; in fact, to all intents and purposes, they do not appear to have young. In the English language, terms which describe many marine creatures have acquired pejorative meaning when applied to humans: cold-blooded, poor fish, spineless, wet, jellyfish, slimy, and slippery are a few examples.

MANAGEMENT IMPLICATIONS

The nature of many processes and problems in the marine environment may appear similar to those encountered in the more familiar terrestrial environment. Nevertheless, the scale, the interconnected nature, and the unfamiliarity of marine systems make many of the management issues very different from those of most terrestrial areas. Acquisition of information for planning and management is considerably more difficult, more expensive and more time consuming than in terrestrial environments.

Known management problems are widespread. Most are caused by the impacts of badly managed human activities, particularly pollution, alienation, and overexploitation in coastal and nearshore waters. These problems are amplified by increased human populations, new applications of technology in the use of marine resources and consequent increased demands. Management problems tend to become evident by gradual reduction of environmental quality or loss of opportunity for humans to use and enjoy aspects of marine environments in areas that are accessible. Dramatic issues of global species endangeredness or extinction are rarely substantial factors in marine environment management and are unlikely to be so for species with planktonic larvae.

Because of the scale and linkages of marine environments, their conservation is more clearly a matter of broad-based management of human uses and impacts than is the case on land. Marine environment management has many similarities with the management of wildlife. The major aspect of any natural environment that can be managed is the impact of human activity. By intention or ignorance, human action can rapidly degrade or destroy the natural environment and its capacity to sustain the long-term economic, cultural and scientific needs of human society. The management of plants and animals, and their environments, involves the management of people.

Clearly the basic requirement for marine environment and resource protection is the management of human use and impacts in very large areas. This generates two requirements. The first is for understanding of uses: the historic and current extent and impact of each use; the extent and likely impact of future use on the basis of user expectations; the interactions of the range of present and likely future uses; the apparent capacity of the system to sustain use; and the options for management of each use. The second is to persuade users that their long-term interests will be served by management of the marine environment. If the people closest to the marine environment do not or cannot economically afford to accept the need for management, it will either fail or be extremely costly to enforce.

Chapter 4
The Nature Of Management Problems

Human activities are part of the dynamics of marine ecosystems. The image of coastal fishing villages with human populations in balance with their surroundings suggests a harmony in which human impacts do not exceed or distort the capacity of the supporting marine ecosystem. The image of major coastal cities with supporting industry, agriculture, and industrial scale fishing suggests an accumulation of impacts that may exceed the capacity of the marine ecosystem to sustain itself or to repair itself in the wake of natural or human impacts. Almost all management of the environment or of natural resources consists of limiting or eliminating human impacts.

In the simplest terms, human impacts arise through altering the pre-existing natural system by removing biological or physical resources or by introducing physical, chemical or biological factors that distort processes that maintain the system.

Human impacts on marine systems fall into one or more of the following categories: fishing or collecting—removing or killing species; physical extraction—dredging or mining; pollution—introducing physical or chemical factors that alter the functioning of the system, and; alienation —conversion of areas by constructing or placing physical structures that cover or shade the seabed and shallow waters, alter current flow, or otherwise displace the pre-existing system.

THREATS AND IMPACTS

There are three forms of environmental impact that may be addressed by conservation planning and management. They arise from human impacts upon the structure, processes, and amenity of ecosystems. Structural impacts result from deliberate gross modification or destruction of ecosystems in order to make different use of the areas in which they occur. Process impacts arise from alteration of physical, chemical or biological factors as a result of indirect, incidental and often unintended impacts. The consequence of process impacts may be structural change. Amenity impacts are the result of reduction of the range of current and future options for use of natural areas and resources for a wide range of purposes including those not currently anticipated.

On land, structural conservation has often been the most obvious, emotive and urgent problem. The clearing of preexisting ecosystems to make way for agriculture, human settlements, industry, mining, or transport corridors generally causes immediate structural impact. Further structural impact may also occur less immediately as a consequence of destroying or damaging physical or biological processes by fundamental engineering modifications such as damming catchments or diverting flood channels that may destroy or disrupt habitats, migration routes, or other natural linkages. Structural impacts present readily demonstrable and often profound threats to many terrestrial ecosystems and endemic species.

Structural damage to marine ecosystems occurs particularly in the shallow coastal fringes where it takes the form of dredging, port development, coastal stabilization, causeway construction, mariculture pond development and land creation through alienation or reclamation. As a coast becomes heavily developed the increasing extent of alienation of particular community types may become a matter of structural conservation significance. An example is the destruction of mangrove on tropical coasts.[1]

On land the protection of the structure of precisely definable areas of habitat essential to the survival of endangered species has been a major priority. In marine ecosystems structural damage has generally been a less substantial issue. Local impacts may be severe but because of the degree of linkage within and between marine environments, a specific development can rarely be identified as a terminal threat to a particular species. Sensitive sites exist, particularly in estuarine and nearshore areas, but not to the extent that species extinction of non-airbreathing marine species has been identified as a significant issue.

The second class of impact is process damage or distortion. This involves indirect, incidental, and generally unintended effects on ecosystems through alteration of physical, chemical or biological factors. Concerns to date generally relate to process damage or distortion resulting from human pressure through over exploitation of resources and pollution. In general, the consequences of inadequate process conservation are not immediately obvious and tend to take the form of gradual decline in environmental diversity and resilience or capacity to recover from impacts, particularly in areas most accessible to people.

In the sea, there are many examples of the decline of exploited resources: the tragedy of the commons;[2] and some dramatic examples of pollution such as oil-covered marine mammals or sea birds, tarballs and floating plastic in mid ocean areas.[3] Some process threats are addressed by regulation of fisheries, others are being addressed by a growing number of international conventions on pollution and shipping[4] Generally, the symptoms of inadequate process conservation are not obvious or dramatic. Some issues, such as the accumulation of heavy metal in fish and shellfish tissues to levels that make them toxic to humans, may become apparent in dramatic circumstances but require major scientific programs to identify their cause.[5] Given the nature, extent and degree of linkage in most marine ecosystems, process conservation generally involves large-scale systemwide planning and management.

The third class of impact concerns the maintenance of amenity value or the options for human use of natural areas and resources. Management to maintain amenity involves considering present and future use and potential use of natural environments including uses and purposes not yet known. It involves seeking economic and environmental plans based upon conscious choices regarding the nature of the interaction between human populations and sustaining a productive, diverse natural environment.[6]

The nature of amenity value covers a broad range of human interactions with the marine environment. The most common, at one extreme, is the utilitarian approach of value in terms of materials collected or harvested for food or as materials for construction or ornament. At the other extreme are cultural, spiritual, or philosophical views of the value of wilderness, undisturbed by humans.

In the absence of clear identification of critical areas necessary for structural conservation and in light of the paucity of knowledge concerning process conservation, amenity conservation becomes an important management tool. This is particularly true in accessible areas that are culturally important to local communities.

MANAGEMENT APPROACHES

Conceptually, the logic of management planing is simple. It consists of problem definition—identifying the impacts related to each activity, and problem rectification—incorporating suitable controls and limitations in a management plan.

It is rare for management to have the luxury of dealing with a simple situation where the issues relate to a single impact and a single impacting group. The environmental manager must therefore aspire to contain perceived or probable impacts by controlling the method and the extent of each relevant human activity and by influencing its motivation.

Typically, as it expands, an activity passes from being reasonable and sustainable with no apparently significant impact to having a level of impact that is regarded as acceptable and finally may reach the state where it is recognized as causing unacceptable environmental consequences. At that point there are clear issues that may be defined in terms of substantial threats to structure, process or amenity.

In marine environments the task of environmental management problem definition occurs in a highly dynamic setting. The methods and the extent of existing uses and impacts change and new uses and impacts develop. This is further compounded by lack of understanding of the dynamics of the complex physical/biological processes that shape marine communities. Major changes may be observed, but it is often difficult or impossible to determine whether they are the consequence of natural dynamics, or of human impact affecting the structure or processes of marine environments. If human impact is considered to be a possible cause there is then the matter of determining whether the observed affect results from the direct impact of a single human activity or from an interaction of two or more activities altering natural dynamics.

An example is the phenomenon of crown of thorns starfish. Large populations of this starfish have occurred twice in the last 20 years on coral reefs throughout the Indo-Pacific region. In the course of about two years such a population can, by its predation, kill most of the coral on a reef. A reef that has been affected is not aesthetically appealing. With the death of corals there is a loss of the attractive fish associated with coral communities. Over time there is a sequence of algal colonization and apparent abundance of grazing fish species, and eventually coral cover is restored through regeneration of colonies from fragments unaffected by the starfish, or from recolonisation by newly settled larvae. Coral cover may be restored within 20 years but it may take much longer for species diversity and age/size structures to resemble those that occurred at a site before starfish predation. On the Great Barrier Reef, there is sedimentary evidence to suggest that the starfish has been a factor in reef environments for thousands of years. Despite this, there is not sufficient evidence to determine whether populations have occurred before with the density and frequency of those observed recently.[7] Neither, after two decades of intensive research, is there sufficient evidence to support or dismiss hypotheses that large populations are the consequence of human intervention through pollution or predator removal. There is consequently insufficient evidence to determine the extent to which the problem of the crown of thorns starfish is one of structure, process, or amenity.

If the starfish is a normal component of reef dynamics, occasionally occurring in large numbers since before the time of possible impacts from industrialized human society, the problem is largely one of amenity. If a reef is not aesthetically appealing and if its fish population is changed, its amenity value for recreation, tourism and possibly fishing is reduced.

If crown of thorns starfish occurrences have become more frequent, they may represent threats to the structure and the process of coral reefs, changing the species balance and perhaps affecting the rate of accumulation of calcium carbonate.

The management options are limited. Starfish numbers can be controlled at great expense at individual sites, but no method exists for widespread control. Management can seek to address the issue of amenity immediately by removing starfish at sites of high amenity value. Management can support research to determine the cause and impact of large populations of the starfish.[8]

A consequence of mixing and linkages in marine environments is that almost all management situations are likely to be similarly complex through interactions between the impacts of several human activities and the dynamics of a little known system.

Determination of the progression from acceptable to unacceptable impact involves social and economic as well as ecological judgment. The need for costly restrictive measures to address impacts is generally not accepted until either the causal link has been demonstrated with a high degree of statistical confidence or the problems are so advanced that users are desperate for anything to be tried. It is thus important to understand the social and economic motivation of activities in order to develop an effective approach to solving the environmental problem. The management remedy depends ultimately on being able to develop a community motivation to remove or so greatly reduce an impact as to make it insignificant.

CONSERVATION PERSPECTIVES

The components of most marine environment management problems have been described as a consequence of the results of research and management experience in many parts of the world. There is now a considerable literature documenting ways in which particular human activities can modify, disrupt, or destroy marine environments and the biological and human communities that they support. Nevertheless much of this literature takes the form of descriptive articles published in local journals or government reports not readily accessible to computer-indexed search systems. Often, important material is presented in a complex way, so surrounded by proper cautious scientific qualifications that its significance is not readily apparent to planners or decision makers. Recent initiatives have resulted in the synthesis of such literature in terms more suited to decision makers who are not scientifically trained. Thus, Salvat et al.[9] have provided a solid synthesis of established, probable, and possible human impacts on coral reefs.

A few impacts are immediate, predictable, and locally demonstrable such as those that follow a deliberate decision to alienate and smother an area of seabed to create land. Most are more complex because at low levels, or in a limited number of sites, they may have insignificant effects on structure, process, or amenity, although at high levels they may have profound impacts. This is the case with the loss of mangroves in tropical and subtropical regions.[10] The problems of identifying and analyzing impacts is increasingly compounded by the development of new forms of use creating multiple and often incompatible demands on limited resources.

A long-established class of removal impacts relates to fishing, collecting, or harvesting renewable resources. It is well known that overharvesting will lead to population collapse of the target species. Despite this, the sad conclusion from Berkes[11] is that a common stock, that is, a stock not in effect owned by a human community or individual, will be overfished. Increasing effort will be applied to the stock until the sustainable capacity is exceeded. If the fishery is a matter of subsistence where there is no alternative source of dietary protein, there will be continued and increasing application of effort until the collapse is so complete that the stock is destroyed.

Munro et al.[12] have pointed out that where fishing communities are faced with a desperate fight for survival, the living resources are likely to be grossly overexploited by subsistence fishing. In such a situation, as the catch with conventional methods declines, more and more destructive techniques are used in order to achieve some return per unit effort. In the case of coral reef fisheries, dynamite,[13] poisons,[14] and other destructive techniques such as *muro ami* can all cause major environmental damage. In the muro ami technique, weighted sticks are pounded upon a coral reef in order to scare fish into a set net.

The dynamics of an "export" commercial fishery, taking fish primarily for sale or trade outside the immediate social or economic community that conducts the fishery are more complex.[15] In theory, such a fishery will not continue once the costs of taking fish exceed the economic benefits of selling them. In practice, economic distortions are often such that it may, in the short term, be "economic" to fish stocks to the point of collapse. Where a shallow sea fishery is superimposed upon an existing subsistence fishery displacing any pre-existing system of ownership or control of access to stocks, the tragedy of the commons is a likely outcome. In more remote waters, the efficiency, capacity, mobility and the great operational range of modern fishing fleets have created the commercial logic of "pulse fishing." This involves fishing a stock to levels that can not thereafter sustain fishing for some years and then moving to repeat the exercise at other distant grounds. The immediate commercial success of such strategies has generated a desire to emulate them.

A second class of removal impact concerns capital extraction—the utilization of non- or extremely slowly renewable resources. Examples are dredging of sand or gravel, or mining of the seabed. The impacts depend upon a number of factors. Salvat[16] has discussed the impacts on coral reefs that are particularly vulnerable. The most immediate impacts are removal of the substrate, and with it the attached or dependent plant and animal communities. Depending on the scarcity of the mined substrate, the completeness of the mining activity, and the extent of associated pollution it may result in significant long-term habitat destruction for plant and animal communities. Other long-term impacts may arise from changes to the flow of tidal and other currents as a consequence of removal of previous obstructions. Dredging will often result in incidental or unintended impacts by introducing sediments that reduce light penetration into surface waters and smother substrates. Incidental smothering as a consequence of settlement of sediments introduced to the water mass upstream from the affected area may be gross, but alternatively may be less immediately obvious, although this may still cause substantial changes to the biota of areas by, for example, changing the proportion of fine materials in sediments.

The first form of introduction impact concerns deliberate smothering where areas are alienated or reclaimed to create land, harbors, or causeways. It causes predictable but cumulatively significant losses of important coastal marine habitats such as saltmarshes, mangroves, seagrass beds, or sand flats.

The second form of introduction impact is pollution—the introduction of physical, chemical or biological factors which change the dynamics of an affected area. An annotated list of pollutants follows, indicating their possible effects on marine biota. The list is an expansion by a Great Barrier Reef Marine Park Authority workshop of an earlier table by Salvat and Kenchington 1982.[17]

Some pollutant effects on marine biota are:

1. Herbicides:

   - May interfere with basic food chain processes by destroying or damaging zooxanthellae in coral, free living phytoplankton, algal, or seagrass communities.
   - Can have serious effects even at very low concentrations

2. Pesticides:

   - May selectively destroy or damage elements of zooplankton or ben-

thic communities; planktonic larvae are particularly vulnerable
- May through accumulation in animal tissues have effects on physiological processes such as growth, reproduction, and metabolism
- May cause immediate or delayed death of vulnerable species

3. Antifouling paints and agents:

- May selectively destroy or damage elements of zooplankton or benthic communities
- Likely to be a significant factor in harbors, near shipping lanes, and in enclosed, poorly mixed areas with heavy recreational boat use.

4. Sediments and turbidity:

- May smother substrate
- May smother and exceed the clearing capacity of benthic animals, particularly filter feeders
- Reduce light penetration, likely to alter vertical distribution of plants and animals in shallow communities such as coral reefs
- May adsorb and transport other pollutants

5. Petroleum hydrocarbons:

- A wide range of damaging effects depending upon type of hydrocarbon, dilution, weathering, dispersion, emulsification or interaction with seawater or other chemicals
- Direct contact with living tissue usually results in local necrosis and, with longer exposure, death
- Exposure to water-soluble hydrocarbons results in mucus production, abnormal feeding, changes to a wide range of physiological functions, and, with longer exposure, death
- Detrimental effects on reproduction and dispersion; premature discharge of larvae, distorted larvae, decreased larval viability
- Residual hydrocarbons in substrates may lead settling larvae to avoid affected areas, and thus block recolonization and repair

6. Sewage-detergent phosphates:

- Inhibit a wide range of physiological processes and increase vulnerability of affected biota to a range of natural and human induced impacts
- Inhibit calcification, e. g., in corals and coralline algae
- Can cause effects at very low levels

7. Sewage and fertilisers—Nitrogen:

- Increased primary production in phytoplankton and benthic algae distorts competitive and predator/prey interactions in biological communities in areas such as coral reefs, which are characterized by very low natural nitrogen levels
- Reduced light penetration through absorption and turbidity of increased planktonic communities
- Increased sedimentation of detritus from planktonic communities
- Increased nutrient levels in benthos from sedimentary organic material
- Selectively favors growth of some filter or detritus feeders such as sponges and some holothurians
- Some species such as corals are affected at very low levels

8. High or low salinity water—freshwater runoff, effluents:

- Low salinity water floats on top of water column, high salinity water sinks, prior to mixing and dispersion
- Tolerance of species highly variable so changed regime may alter biological communities, particularly those in shallow, poorly mixed or enclosed waters
- Salinity is a key factor in settlement and physiological performance of many shallow benthic and reef organisms
- May (e.g., for corals) cause physiological stress evidenced by elevated mucus production, expulsion of zooxanthellae, or death

9. High or low temperature water—from industrial plant heating or cooling:

- Tolerance of species highly variable so changed regime may alter biological communities, particularly those in shallow, poorly mixed, or enclosed waters
- Temperature is a key factor in settlement and physiological performance of many shallow benthic and reef organisms
- May (e.g., for corals) cause physiological stress evidenced by elevated mucus production, expulsion of zooxanthellae, or death

10. Heavy metals e.g., Mercury, Cadmium:

- May be accumulated by, and have severe effects upon, filter feeders and, by accumulation up the food chain, pass these effects to higher predators

- Can interfere with physiological processes such as the deposition of calcium in skeletal tissue
- May (e.g., for corals) cause physiological stress evidenced by elevated mucus production, expulsion of zooxanthellae, or death

11. Surfactants and dispersants:

- Most are toxic to marine biota
- Synergistic effects of dispersant/hydrocarbon mixes can be more toxic than either component unmixed
- Can interfere with a wide range of physiological processes, e.g., photosynthesis

12. Chlorine

- At low levels inhibits external fertilization of some invertebraes, e.g., sea urchins
- Can be lethal to many species

DIFFERENT PERSPECTIVES ON MARINE CONSERVATION ISSUES

Perspectives on conservation depend upon personal judgments regarding the amenity value of the environment. This covers a broad range of intentional human interactions with biological resources of natural areas. For those who have a direct interest in maintaining the nature and productivity of the marine environment, there appear to be three approaches that are not mutually exclusive. The first is the long-standing utilitarian approach of value in terms of materials collected or harvested for food, ornament, manufacture, or construction. The second is the philosophical approach, developed from cultural or spiritual beliefs or from scientific principles, which places value on the maintenance of areas essentially undisturbed by the intrusion of human activities. The third is a more recently recognized utilitarian approach that places value on essentially natural areas for recreation, tourism, and education. Multiple use management approaches may be summarized in a diagram developed from Kenchington[18] which illustrates the area in which a decision or group of decisions will conform technically with a requirement to address the concerns of the three interest groups, with the "perfect" solution represented by the midpoint of the triangle (Figure 4.1).

Discussion of amenity value often identifies profound differences of view over the nature and desirability of resource use and the relationship between humans and other elements of the natural environment. These differences often reflect the degree and nature of economic dependence upon the resources of natural areas. The more amenity value relates to philosophical issues or psychological perceptions, the further the decision-making process departs from one to which quantification can make decisive contributions. Physical or biological factors can be measured or socioeconomic units of currency or employment creation can be derived by each interest group to support the case for a solution that meets its objectives. Nevertheless,it is rare for such a case to be so conclusive that it can be generally accepted as objectively correct. Biological resource decision making should be recognized overtly as an issue of cultural and political choice in which sustainability and the range of benefits to human society as a whole and the relative benefits to individuals and groups within that society are key factors.

Many issues involve structure, process and amenity. Thus competition leading to excessive harvesting of a biological resource may cause structural impacts that become increasingly apparent as the resource becomes scarce. It may cause process impacts depending upon the role of the harvested species in the food web and the extent to which other species will be affected by its decline. It will cause amenity impacts to those who have depended on the resource species or other affected attributes of the environment in which it occurs. It will cause wider amenity impacts if the structural and process impacts have affected fundamental preservation or recreation options.

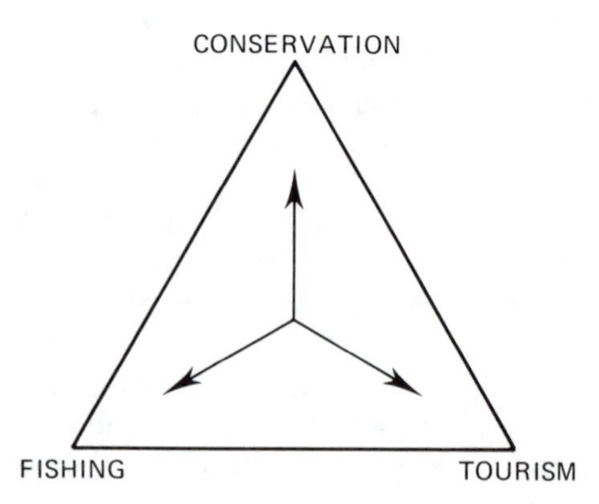

Figure 4.1 Diagram illustrating the three user interest groups which must be taken into account in developing a multiple use managment plan for a marine area.

Effective amenity conservation planning thus requires a "bottom-up" approach building from two bases: the ecological basis of the best available understanding of the natural systems and processes of the area; and the socioeconomic basis of the needs and expectations of those who use or value the resources of the area. This often involves research and community education in order to demonstrate cause and effect of human-impacts and to establish that management has the potential to halt or reverse decline in amenity.

The nature of amenity can change rapidly. Factors leading to change include growth in human population, economic development, improved technologies, and greater mobility resulting from improved transport. In many areas these have resulted in rapid transition where new uses and values have been superimposed upon long-established use by local communities valuing marine resources for subsistence and artisanal fishing. Typically changes flow from increased population and increased economic expectation. Part of the community may develop commercial fisheries that derive economic benefit from fish sold outside the local community. Another part may concentrate on economic activities that are not directly concerned with the quality or productivity of the marine environment and may thus value the amenity of the coastal sea as a means of disposal of wastes. Improved transport makes such areas more accessible to markets and consequently to accelerating economic development and eventually to recreational use and the development of commercialized recreation or tourism.

The economic value of a fish as an item of food is diversified. First it becomes the monetary value of the fish sold in the market, next of the amount a person will pay for the recreational pleasure of catching it, in lieu of the pleasure of seeing it alive, and finally the amount a person will pay for the knowledge that it is alive in its natural environment. This is more a cumulative process than a simple progression. As each new approach to use and value is introduced, it competes with those that existed before.

The issues and problems of regulation of fishing and shipping and coastal engineering are compounded by competition with alternative forms of use of natural resources, such as tourism or nature appreciation. They are further compounded by the pollution effects of other activities whose primary purpose is not dependent upon or affected by the quality or productive capacity of the marine environment. A consequence of the strength and length of linkages in marine environments is that many of the impacts are felt at a great distance from the point at which they originate. The causer may well be unaware of the impact. If the impact does not take the form of toxic seafood, floating dead animals and debris, discolored or odorous water, it is unlikely to be widely noticed and even less likely to be readily associated with its cause.

A distinction can be drawn[19] between direct and indirect users in terms of likely interest in and support for management measures that restrict an activity in order to safeguard, maintain or enhance environmental quality.

Direct users are those whose primary purpose depends directly upon the quality and productive capacity of the marine environment. They include fisheries and some forms of recreation and tourism. They benefit from and are affected by the maintenance of environmental quality. Unless they are suffering from serious short-term economic problems, they are likely to be concerned about it. They are likely to accept, at least in the long term, management measures that maintain the productive natural qualities of a marine environment.

Indirect users are those whose primary purpose is not directly related to the quality and productive capacity of the marine environment. They include coastal development, waste disposal, and transportation. They are not apparently directly affected by, or necessarily concerned about, decline in marine environmental quality. They are likely to consider themselves disadvantaged by additional costs imposed upon their activities in order to safeguard environmental quality.

THE NEED FOR COORDINATION

Individual activities involving direct or indirect use of marine environments have tended to be managed separately with informal or ad hoc mechanisms for resolving issues between sectors. In coastal areas there are often two and sometimes three levels of government jurisdiction with dozens of agencies and authorities involved. The diversity of agencies, many of which are not likely to regard conservation of the marine environment as a primary or even significant goal, makes the establishment of effective coordination to focus on marine environment management at an appropriate scale an extremely challenging task. The fragmentary approach to management of activities that affect the marine environment is well illustrated in two tables taken from Sorensen et al.[20] In table 4.1 those sectors listed by Sorensen et al. that correspond to direct uses in the terminology of Kenchington[21] are identified by an asterisk. Almost two-thirds of the sectors identified by Sorensen et al. are indirect users for whom costs of management to retain, conserve or restore the marine environment are likely to represent unproductive on-costs. Table 4.2 illustrates the range of government agencies at federal, state and local government levels that are involved in the regulation of five selected sectors of marine activity.

PROBLEMS OF INTEGRATING A RANGE OF USES OR IMPACTS

For direct users, management requires control of demand and impact so that they do not exceed the supply or natural regenerative capacity of the marine environment. Issues may be addressed through activity-specific legislation and agencies. For long-established activities there is generally a need to devise strategies to address internal competition, arising from increases in the extent and efficiency of the existing use. There may also be a need to address external competition arising from the introduction of newly developing forms of use that exploit or affect aspects of the target resource. With a resource base limited by a finite ecological capacity, management is likely to involve difficult economic and political decisions between sectors.

For indirect users, management requires coordination. The object of management should be to achieve the minimum practicable usage impact upon the ecological processes, natural resources and options for sustainable direct use of the marine environment. The costs imposed on a user to

Table 4.1

Sectoral Planning and Development in the Coastal Zone

Sectors that are often coastal zone-or ocean-specific	Sectors that are rarely coastal zone-specific but have direct impacts
1. Navy and other national defense operations (e.g. testing, Coast Guard, Customs)	1. Agriculture
2. Port and Harbor development (including shipping channels)	2. Forestry
3. Shipping and navigation	3. Fish and wildlife*
4. Recreational boating	4. Parks and recreation*
5. Commercial and recreational fishing*	5. Education*
6. Mariculture*	6. Public health-mosquito control and food
7. Tourism*	7. Housing
8. Marine and coastal research*	8. Water pollution control
9. Shoreline erosion control	9. Water supply
	10. Transportation
	11. Flood control
	12. Oil and gas development
	13. Mining
	14. Industrial development
	15. Energy generation

* Activities that have a direct interest in marine environmental quality.

Table 4.2

Arrangement of Government Organizations in the United States for five Selected Sectors of Marine Activity*

Functions	Port Development	Fisheries	Pollution Control	Parks & Recreation	Marine Research
Policy Setting and Plan Making	CG COE LPD	NMFS SD SFGD	EPA COE CG	NPS FS LG	NSF NOAA ONR
Regulation (Permit Letting)	CG COE LPD	NMFS CG SFGD	EPA SWQA COE	NPS LG SPD	NOAA FWS SFGD
Levy Charges	LPD	NMFS SFG	SWQA LG	NPS FS SPD	N/A
Fund and/or Construct Projects and Programs	COE EDA LPD	NMFS EDA SBA	EPA SWQA LG	NPS FS SPD	NSF NOAA ONR
Acquire, Manage and Sell Property	GSA CG LPD	GSA LG	GSA LG	NPS FS LG	GSA UNIV NOAA
Generate and Disseminate Information	UNI V MA	NMFS SFG UNIV	EPA SWQA UNIV	NPS FS UNIV	NSF NOAA ONR

Taxation	IRS STB LPD	IRS SFCD	IRS	IRS STB	Tax Exempt

*Key to Acronyms for government organizations

1. U.S. Federal Agencies

CG	Coast Guard
COE	Army Corps of Engineers
EDA	Economic Development Administration
EPA	Environmental Protection Agency
FS	Forestry Service
GSA	General Services Administration
IRS	Internal Revenue Service
MARAD	Maritime Administration
NMFS	National Marine Fisheries Service
NOAA	National Oceanic and Atmospheric Administration
NPS	National Parks Service
NSF	National Science Foundation
ONR	Office of Naval Research
SBA	Small Business Administration
SD	State Department

2. State Agencies

SFCD	State Fish and Game Department
SPD	State Park Department
STB	State Tax Board
SWQA	State Water Quality Department
UNIV	State Universities

3. Local Government Agencies

LG	Local Government (City and/or County)
LPD	Local Port District

minimize impact and the immediate social and economic benefits of maintained or improved environmental amenity are often separated between levels of government within a nation and sometimes between nations. There is thus usually a need to establish effective coordination between governments at the highest level.

In the long term, for both direct and indirect users, the economic and social consequences of environmental degradation and collapse may be severe. Logically therefore, ecological criteria should take priority over short-term social and economic criteria. Practically, however, longer term management is unlikely to be effective if there is no strategy to address short-term social and economic needs identified by users.

Integrative management should thus address the issues raised by all the individual direct and indirect use sectors. Most are discussed in specific texts, a selection of which is listed in the bibliography. For the purpose of this overview a brief coverage of the issues raised by a number of uses and impacts is discussed. The reader is urged to refer to more substantial sector-specific texts for more detailed analysis.[22]

In qualitative terms, issues of environmental conservation may be perceived by users as opportunities or threats to their amenity, activities or values. One approach to this is the development of user profiles and discussing them with the user groups. Four examples are given below of some generalized user profiles for activities in the shallow tropical Pacific from Kenchington (in press).[23] An opportunity threat profile is often an important first step toward identifying the critical issues and obstacles to the development of sector-specific or coordinated management. The example in the following list is developed from Kenchington.[24]

Sample opportunity threat profiles for some uses of shallow tropical Pacific Ocean environments are:

---

1. Recreational subsistence fishing and collecting:

   a) Threatened by
      - increased human population leading to overfishing
      - increased participation leading to overfishing
      - local competition from "market" and "export fishing"
      - distant competition and impacts on stocks of migratory species

   b) Socioeconomic factors
      - often an important source of human dietary protein
      - often closely associated with leisure/tourist interests
      - may be amenable to catch limitation controls

- depending on impact, may be consistent with protected area management

c) Main information needs
- biology and ecology of target species
- impact on populations and habitats of target and non-target species
- importance to local and national economies
- extent of participation and projected growth
- options for reducing impact

2. Export commercial fishing:

a) Threatened by
- overfishing
- introduction of highly efficient technology
- site competition from other activities
- damage, loss or destruction of habitat of target species

b) Socioeconomic factors
- likely to resist management until problems are severe
- tends to become over-capitalised and therefore economically vulnerable in the short term to restrictive measures introduced to provide for long-term sustainability

c)Main information needs
- biology and ecology of target species
- impact on target and non-target species
- level of activity and projected growth
- options for reducing impact
- importance to local and national economies

3. Visual/photographic tourism and recreation:

a) Threatened by
- damage, degradation or destruction of habitat
- impact of severe natural events such as cyclones
- site competition from other activities

b) Socioeconomic factors
- likely to derive immediate economic benefit from measures to protect the environment
- economic importance in comparison to fishing and collecting

activities is often overlooked or underestimated
- generally compatible with protected area management

c) Main information needs
- impact on the environment
- level of participation and projected growth
- options for reducing impact
- importance to local and national economies
- alternatives

4. Representative environment preservation and research:

a) Threatened by
- damage, degradation or destruction of habitat
- increasing demand for other uses

b) Socioeconomic factors
- essential component of protected area management
- may be achieved within highly protected sub-units of areas set aside for visual/photographic recreation and tourism

c) Main information needs
- understanding of the character of the environment sufficient to identify what should be represented
- identification of the linkages affecting representative areas of habitat
- impact upon established human activities
- importance to local and national economies
- alternatives

The lack of clear boundaries, the diffuse nature of many forms of use or impact, and dynamic changes in the ease of access to, and forms of use of, marine resources create major challenges in the development of effective management strategies. The issues—preservation of representative examples of ecosystems, conservation of renewable resources, and provision of a measure of equitability in allocating reasonable use—are the same as those faced by terrestrial managers but the nature and scale of the specific problems are generally different.

The task of developing management solutions is complicated by the cost and logistic problems of collecting data in marine environments discussed in the previous chapter. These problems and the critical importance of socioeconomic factors force the planner and manager to work with users in the assembly of information for management decisions to an extent that is unusual on land.

Chapter 5

Establishing A Framework For Managment

Converting an awareness of need into action to manage marine areas or resources generally involves detailed consultation and planning. Such a process is futile if it is not matched by a long-term commitment of people, equipment, finances and resolve to implement the plan, monitor the outcome and ensure that the exercise is more than just a matter of creating plans or statements of intent.

The first step is to develop the awareness of an informed or concerned minority into adoption of a series of goals or objectives that can be appreciated and adopted through community or government action. Those goals may be addressed by a single coordinating marine management plan or by looser interactions between the sectoral plans of competing agencies. The approach will depend upon the socioeconomic structure and decision-making processes of the communities involved. These may range from traditional management decision-making practices[1], to the planning processes of technologically complex societies.

The actual goals or objectives are likely to be substantially similar across a wide spectrum of situations. The components will typically include achieving maximum sustainable economic benefit from long-term use of natural resources, maintenance of the essential nature and natural productive capacity of an area, and allocation of resources between competing users or uses. They are increasingly likely to include establishment of some protected areas for purposes of recreation and tourism and others as reference sites for research and as sanctuaries.[2]

The way in which such goals are translated into actions again depends on the socio-economic structure and decision making processes of the human societies involved. It may involve executive decree or creation of legislation by a parliament. A key decision is often whether to seek coordination of the wide range of activities and policies by the creation of new legislation and a new organization or by extension and specified interaction of existing legislation and agencies such as those involved in fisheries, coastal shipping and management of waste discharge.

THE LEGISLATIVE BASE

Kelleher and Lausche[3] produced a set of guidelines for the development of legislation for marine resource management. These guidelines were deliberately expressed in general terms so that they might be applied to a wide range of legislative and executive systems. They have been refined and the most recent version,[4] is presented here:

Need For A Policy

An overall policy on the management, use, and conservation of marine and estuarine areas should be developed for a nation as a whole, for regions of the nation, where appropriate, and for any identified sites of particular significance at the national level. Ideally such a policy should address coordination with management of coastal lands and catchments. The process of creating the policy, as well as its existence and provisions, will contribute to national recognition of the importance of conservation and management of marine and estuarine areas, to the selection and establishment of an appropriate system of Marine Protected Areas and to attainment of a primary goal of management—sustainable use. The policy may be established within a national or regional conservation strategy.

Statement of Objectives

Objectives encompassing conservation, recreation, education, and scientific research should be written into legislation. If this is not done and if conservation is not given precedence, setting aside of protected areas may be an empty political gesture. A primary conservation objective in resource management legislation must be recognized as essential to sustained use and enjoyment of the resource.

Linkages Between Marine Environments

The young of many marine animals and their food, as well as plant seeds, propagules, and pollutants, are transported in the water column, often over distances so great that they cover the territorial waters of several countries. Marine animals also often migrate over great distances. To the greatest extent achievable, legislation and policy should shape and take into account regional, international and other multilateral treaties or obligations. Such an approach attempts to ensure that the management initiatives of one country are not negated by the actions of others connected through the transport of recruits, food or pollutants, or through the migration of marine animals.

Sustainable Use

The legislation should recognize the linkage between protection and maintenance of ecological processes and states, and the sustainable use of living resources. Explicit reference to the objectives and concepts of the World Conservation Strategy[5] may reinforce the legislation and its effectiveness.

Multiple-Use Managed Areas

It is strongly recommended that legislation be based on sustainable multiple-use of substantial managed areas as opposed to isolated highly protected pockets in an area that is otherwise unmanaged or is subject to regulation on a piecemeal or industry basis. Such umbrella legislation can be justified on the grounds of worldwide experience of conventional piecemeal protection of small marine areas alongside conventional fisheries management. This usually leads to overexploitation and collapse, perhaps irreversible, of stocks of exploited species and progressive deterioration of the protected area. In designing umbrella legislation the following goals merit consideration:

- Provide for conservational management over large areas
- Provide for a number of levels of access and of fishing and collecting in different zones within a large area
- Provide for continuing sustainable harvest of food and materials in the majority of a country's marine areas

Coordination

Coordination of planning and management, by all intragovernment, intergovernment and international agencies with statutory responsibilities within areas to be managed, must be provided within the legislation. Provision should be made to define the relative precedence of the various pieces of legislation that may apply to such areas. Because of the interconnectedness of species and habitats in marine ecosystems, the legislation should provide, within protected areas, for control over all marine and estuarine resources of flora, fauna, terrain, and overlying water and air.

Activities External To Marine Protected Areas

Because of the linkages between marine environments and between marine and terrestrial environments, it is important that legislation include provisions for the control of activities that occur outside a Marine Protected Area, which may adversely affect features, natural resources, or activities within the area. Often low or high water marks constitute a jurisdictional boundary, but these are impractical boundaries for species with life cycles or feeding behavior that involves crossing them. Other boundaries exist between Marine Protected Areas and adjacent marine areas. A collaborative and interactive approach between the governments or agencies with adjacent jurisdictions is essential.

Legal Powers

The power to establish any marine protection/conservation management system should be provided by law, with approval and any subsequent amendments to require endorsement by the highest body responsible for such legislative matters in the country concerned. Establishment in this context includes the requirement that the legislation contain enough detail for:

- Proper implementation and compliance
- Delineation of boundaries
- Providing adequate statements of authority and precedence
- Providing infrastructure support and resources to ensure that the necessary tasks can be carried out

Management Arrangements

If management is to succeed, interagency disputes, concerns, obstruction or delay must be minimized. It follows that legislation and management arrangements should grow from existing institutions unless there is overwhelming public and political support for completely new administrative agencies. Therefore:

- Creation of new agencies should be minimized
- Existing agencies and legislation should be involved by interagency agreements wherever practicable

- Existing sustainable uses should be interfered with as little as practicable
- Existing staff and technical resources should be used wherever practicable
- Unnecessary conflict with existing legislation and administration should be avoided
- Where conflict with other legislation and administration is inevitable, precedence should be defined unambiguously.

Consistency With Tradition

The form and content of legislation should be consistent with the legal, institutional and social practices and values of the nations and peoples enacting the legislation:

- Where traditional law and management practices are consistent with the goals and objectives of the legislation, these traditional elements should be drawn upon to the greatest practicable extent. This applies to both the traditional, perhaps unwritten law of aboriginal communities and the more recent traditions of a country or people.
- The customary or accepted ownership and usage rights of a marine area that is to be managed are critical considerations. Legislation should reflect this. There may be public or communal rights as well as private ownership. Customary fishing rights need careful consideration.

Definitions

The definitions and terminology in legislation should use words that reflect, in language clearly understood by those affected, the intentions, goals, objectives, and purposes of the legislation. Terminology is likely to differ from country to country, but where practicable there is some advantage in adhering to standard terminology.

Responsibility

Legislation should identify and establish institutional mechanisms and specific responsibility for management and administration of marine areas. Responsibility, accountability, and capacity should be specified adequately to ensure that the basic goals, objectives, and purposes can be realized. As well as government agencies, local government and administration, traditional village community bodies, individual citizens, clubs, and associations with compatible goals, objectives and responsibilities should be involved in management wherever practicable.

Management And Zoning Plans

Legislation should require that a management plan be prepared for each managed area and should specify constituent elements and essential considerations to be addressed in developing the plan. It is particularly recommended that, where the multiple-use protected area concept is to be applied, legislation should require zoning arrangements to be described in sufficient detail to provide adequate control of activities and protection of resources. The provisions of zoning plans should overide all conflicting legislative provisions, within the constraints of international law.

Public Participation

Involvement and active participation of users of marine environments in the development of legislation in establishing, maintaining, monitoring and implementing management of marine areas is almost always of key importance to the acceptability and success of management. It is highly desirable that the concept of public participation, expressed in terms appropriate to social and government structure, is established in legislation and that the procedures are sufficiently detailed to ensure effective participation.

Accordingly, opportunities should be provided for the public to participate with the planning or management agency in the process of preparing management and zoning plans for marine areas to be managed including: the preparation of the statement of purpose and objectives; the preparation of alternative plan concepts; the preparation of the final plan; and any proposed major changes to the plan.

Preliminary Research And Survey

It is often a mistake to postpone, by legislation or otherwise, the establishment and implementation of marine management until massive research and survey programs have been completed. Often sufficient information already exists to make strategically sound decisions on the boundaries of areas to be managed, and the degree of protection to be provided within them. Postponement of such decisions often leads to increasing pressure on areas under consideration and greater difficulty in making the eventual decision. Provision in the legislation for periodic review of management and zoning plans allows their continual refinement as research information becomes available and as user demands and impacts change.

Monitoring, Research, And Review

The legislation should provide for surveillance in order to determine the extent to which users adhere to the provisions of management, for monitoring to determine the condition of managed ecosystems and their resources, and for research to assist in the development, implementation, and assessment of management. The legislation should provide for periodic review of management and zoning plans in order to incorporate desirable modifications indicated from the results of surveillance, monitoring and research. The processes of, and the degree of public participation in, plan review should be the same as for initial plan development.

Compensation

Consideration should be given, where local rights and practices are firmly established, to arrangements for specific benefit to local inhabitants in terms of employment in management or of compensation for lost rights, because experience has shown[6] that the success of conservation management programs depends critically on the support of local people.

Financial Arrangements

Financial arrangements for management of marine areas should be identified in legislation according to local practice. Consideration should be given to establishing special funds whereby revenue raising from marine management can be applied directly back to the program or to affected local people.

Regulations

Legislation must provide authority for adequate regulations in order that activities can be controlled or, as necessary, prohibited. Three types of regulation may be considered:

- Regulations to enforce a plan
- Interim regulations to provide protection for an area for which a plan is being developed

- External regulations to control activities occurring outside a managed area that may adversely affect features, resources, or activities within the area

Enforcement and Penalties

To be effective, legislation must provide adequate enforcement powers, and duties. These should include:

- Effective penalties for breach of regulations
- Incentives for self-enforcement and regulation by users
- Adequate powers for professional field staff to take effective enforcement action, including pursuit, apprehension, identification, gathering of evidence, confiscation of equipment and evidence, and laying charges in a court of law
- Provisions, where feasible, for local people to reinforce or provide enforcement. This is practicable when the local people can continue their traditional use of the managed area even if limitations on that use have to be applied

Education

To be effective, management should be supported by educational measures to ensure that those affected are aware of their rights and responsibilities under any management plan and that the community supports the goal of the legislation. Few countries could afford the cost of effective enforcement in the presence of a generally hostile public. Conversely, costs of enforcement can be very low where public support exists.

A well-designed education and public involvement program can generate political and public enthusiasm for management, its goal and objectives. The design of the education program and materials should include planning and consultative discussions to ensure presentations of information that will be effective with:

- Local resource users
- Researchers and technical specialists
- Industry lobby groups
- Non-government organizations
- Local government
- Other government agencies

TECHNIQUES AVAILABLE TO MANAGERS

A management plan establishes a framework for controlling human activity. It may take a range of forms from an oral agreement or tradition to a legally drafted document. A plan is prepared in order to achieve a goal or goals for those who authorize it. Authorization may come from society as a whole through a mechanism of constitutional government, or it may come from a sector of society with particular interests in the area covered by the plan. The goals may be clearly specified within the plan or they may be unstated but contained within the agreement, tradition, or document that authorizes the plan.

A manager or planner approaching an area to develop a plan will generally find existing controls or constraints on human activity. These may range from informal understandings by users to formal requirements under a variety of pieces of legislation. The body of general regulations applying anywhere within the maritime jurisdiction of a coastal state constitutes a form of management plan. Usually such regulations have developed over time as a consequence of single issue management measures for activities such as fishing or the discharge of wastes. There is typically limited coordination between regulatory activity under different pieces of legislation and consequently there may be inconsistencies. Thus, levels of permitted discharges into coastal waters may differ between terrestrial outfalls and vessels.

A prerequisite for the introduction of a new form of management is to analyze existing controls in order to understand how they operate, who benefits from them, and to identify gaps, conflicts, deficiencies and practical requirements that must be addressed to meet new objectives. Whether analyzing existing practices or customs, or designing a new plan of management, there is a limited range of techniques that may be applied singly or in combination to control the interaction between human activity and the environment.

PROHIBITION

The simplest form of regulation is absolute prohibition of access to a prescribed area. It is easily understood and easily policed. If a person enters the prescribed area it is clear that an offense has been committed.

Absolute prohibition of access is rare. The managing agency will generally have to make arrangements at least for research and monitoring even in the most highly protected areas. The prohibition thus applies to entry without the specific approval of the managing authority. Presence in the prescribed area is no longer proof of offense.

The next most simple control is prohibition of a class of activity within a prescribed area. It is also easily understood and easily policed although it requires the managing agency to establish that the particular prohibited activity has been conducted in the prescribed area in order to establish that an offense has been committed.

The simplicity of prohibition as a control mechanism is that it establishes a black/white or yes/no basis. Management control consists of determining whether or not the prohibited access or activity occurs.

LIMITATIONS

In reality, few activities are so damaging that total prohibition is the only appropriate form of management. Most, such as fishing, waste disposal, or reclamation, are sustainable or insignificant at low levels but become damaging or unsustainable at high levels or at sensitive locations. The task of management is to protect the structure and processes of marine environments by enforcing measures that limit such activities to levels that can be sustained without causing significant impacts.

Limitations may also be an appropriate approach to management where two activities that are sustainable may interfere with each other's amenity. Examples include interactions between sailing vessels and power boats, commercial and recreational fishing, spearfishing and underwater photography.

Limitations may be applied to restrict activities to sustainable or acceptable levels by controlling the area and time in which the activities may occur, the type and extent of equipment that may be used, or the number and skill of the people who may participate. By a combination of these methods it is possible to design sophisticated regimes that should meet the requirements of conservation and reasonable use of marine areas. The problem of management by limitations is that it is generally much more complex and demanding than simple prohibition. The management regime is harder for users to understand and for managers to enforce. Six general types of limitation strategy are widely used in marine resource and area management.

Spatial Controls

A regulation may apply generally, that is, throughout the maritime jurisdiction of a coastal state, or it may be limited to a part or parts of the area under that jurisdiction. At its simplest, spatial control defines the area of application of a management scheme. More elaborately, it can be used to establish zones and thus to assign uses within a managed area.

A zoning plan will subdivide a managed area into two or more subareas with defined activities allowed in some areas but not in others. This may be achieved either by specifying the permitted uses and stating that other uses are prohibited or by specifying prohibited uses and stating that other uses are permitted. The effect or intent of such a plan can thus be determined by listing the range of activities that can legally take place in the managed area and identifying those that are not allowed.

Temporal Controls

The needs of management are not constant in time. On the shortest time scale there may be a need to prohibit access to an area during part of the 24-hour cycle when it is particularly vulnerable to disturbance or when it is not possible to provide adequate management supervision.

In many situations management needs are not constant throughout an annual cycle, thus many species may be more vulnerable and need higher levels of protection at predictable periods within a year. Such vulnerability generally relates to the reproductive cycle; fish are vulnerable in spawning aggregations, birds and turtles when nesting, and marine mammals while giving birth or caring for newly born young. Migrating animals may be particularly vulnerable at particular sites or phases of the migratory cycle.

In addition, feeding sites along routes and in the areas in which migratory species overwinter or spend the part of the year when they are not involved in reproductive activity may be of critical importance to such species for a few weeks of each year. Such sites may require a high level of protection throughout the year, or it may be adequate to regulate human use only during the time of the year that they are used by migrant species. The Ramsar Convention provides for the protection of coastal wetland sites important to the life cycle of internationally migratory bird species.[7]

On a longer time scale, there may be a need to provide a period of increased protection to an area for a period of one or more years. This may be required as part of a fishery management strategy in which areas are used for a time and then closed for a time to enable a depleted stock to recover. It may similarly be required to allow for recovery after a period of recreational use has led to cumulative impacts on vegetation or corals. It may also be needed to protect an area from human impact while it is recovering from a catastrophe such as a severe storm or pollution event.

Equipment Restrictions

Some forms of equipment or techniques of use may be effective for their immediate purpose but have particularly damaging impacts on the environment. This is particularly the case with a number of fishing techniques. Some, such as the use of dynamite to kill fish, may be so damaging that they should be banned absolutely. Others may be acceptable at low levels but unsustainable at high levels. A large part of the fisheries legislation of most coastal states consists of measures to limit the use of efficient techniques to levels where their impact on the target stock or its environment may be sustainable. Thus there are regulations to set minimum sizes on the mesh of nets in order to allow juvenile fish to escape, there are restrictions on the number or size of traps, nets or vessels that may operate in a particular fishery or area and there are bans or limitations on the use of SCUBA equipment for collection fisheries.

Non-fishery equipment restrictions may be applied to the use of anchors that damage sea grass or corals. Another example is limiting the speed of vessels in sheltered waters where the wake may cause erosion of banks or shorelines.

Quotas

An alternative to banning activities or restricting the area of operation or the type of equipment used is to set limits so that the harvest does not exceed the capacity of target natural resources or the environment. In fisheries management this may be achieved by determining before the start of a fishing season, the amount of harvest to be taken during that season. Once the predetermined harvest has been taken, the fishery closes for the season. Quotas may apply to the fishery as a whole, with fierce competition for available stock until the closure level is reached. Alternatively quotas may be issued to individuals with the possibility that different quota levels may be allocated to different categories of users. In amateur or subsistence fisheries a quota may apply on a daily basis by the application of a bag limit that restricts the number of fish a person may take in a day.

In other management systems, often applied to recreation or tourism, quotas may take the form of allocation of a predetermined carrying capacity designed to protect the structure, process, or amenity of the resource. The determination of carrying capacity is a complex process involving scientific assessment of the amount of use a site can sustain without significant degradation. It also involves socioeconomic consideration of the nature and extent of the experience or activities that should occur within the ecological threshold of that site.

Recreational or tourist quotas may be applied in a number of ways. Thus there may be a limit, at any one time, to the number of people allowed onto a particular beach or island, or to carry out a particular activity such as diving at a particular site. Again, the quota may be applied on a daily or seasonal first-come, first-serve basis or allocated to individuals. For example, first-come, first-serve may be applied by a system whereby moorings are provided and no further boats may enter a site once all the moorings are occupied. Allocated quotas are widely used in the regulation of tourist operations providing security of access to a specific site or group of sites.

Skill Licenses

Most fisheries and boating management schemes have some system whereby individuals are required to demonstrate knowledge and technical competence before they are allowed to operate without the supervision of someone experienced in fishing and boat operation. This may be achieved through an apprenticeship scheme whereby young people are trained by assisting family or other recognized local experts. In other systems it may involve theoretical or classroom studies as well as practical experience. Licensing is becoming increasingly important in marine recreation and tourist activities, such as SCUBA diving or speedboat driving, where incompetent operators may cause considerable public risk. Skill licensing, particularly if accompanied by the possibility of withdrawal for non-compliance, can be used to establish codes of practice to minimize adverse environmental and social impacts of an activity. Knowledge of management regulations and the way in which they apply in areas relevant to the individual can be made a compulsory part of a training program leading to a skill license. Failure, particularly repeated failure, to abide by management regulations can then be grounds for loss of license.

Resource Allocation Licenses

A skill license demonstrates that an individual is regarded as competent to undertake a particular activity. In situations where resources are limited, or where there are social or economic reasons to limit access to a resource, quotas may be issued through the allocation of licenses. This enables the manager to limit the impact of activities to the capacity of the resource by limiting the number of licenses allocated.

PERMITS

A permit system provides scope for management discretion in establishing the conditions under which an activity may occur. It can be extremely demanding of management resources because each application has to be treated on a case-by-case basis and there is generally an expectation that this will be done in a reasonably short period. The advantage of a permit system is that it enables managers to assess and make specific provision for the likely effects of particular and precisely described activities at specified sites. A permit system can provide a level of detailed consideration that is impracticable in general plans.

Within a multiple-use plan, permits may be used to allocate quotas and monitor activities the extent and impact of which were not adequately known at the time of planning. These may be new activities or they may be pre-existing activities that have not previously been studied, where there are grounds for concern that they may have significant impacts or that they may grow beyond sustainable capacity.

A major application of permit systems is in the assessment and establishment of conditions for substantial developments. Within the permit process there may be requirements for environmental impact statements to predict the likely impacts and consider alternatives to minimize impacts on the environment and on the amenity of other users. In issuing a permit the management agency may impose conditions on design, form of construction, and operational procedures.

To summarize, the provisions of a management plan establish purposes for which areas may be used or entered.

- A use or purpose of entry may be "of right", that is, any person may undertake that use or purpose of entry without restriction. Such a use is generally also subject to regulations under other laws where compliance with those regulations is not inconsistent with the provisions of the zoning plan.
- A use or purpose of entry may be "of right but subject to any condition specified in the plan." Conditions may include skill or quota licensing.
- A use or purpose of entry may be allowed only after prior notification.
- A use or purpose of entry may be allowed only with a permit.
- A use or purpose of entry not specified as of right. after notification or by permit may be allowed by a permit if the applicant can demonstrate to the management agency the proposed use is "consistent with the management objectives of the area".

Most plans contain a mixture of techniques for the regulation of use. They may be used in complex combinations. Thus a fishery may be restricted to licensed master fishermen who also hold a quota or access license. It is likely also to have a closed season, gear or equipment restrictions and to have a number of areas that are off limits as known nursery areas, national parks, or recreational reserves. Often, complex systems have arisen by cumulative addition of restrictions as the limits to an activity have become apparent. Often, there is scope to simplify regulation without restricting the management objectives.

Chapter 6

Developing A Management Plan

The planning approach discussed here is developed from that outlined in the UNESCO Coral Reef Management Handbook[1] which was in turn developed from processes evolved in management planning for the Great Barrier Reef Marine Park.[2]

THE PLANNING PROGRAM

The minimum requirements of the planning program will be determined by the legislation providing for management plan development. The program need not necessarily be restricted to the minimum and ideally it will have five phases:

1. Initial or premanagement information gathering and preparation. The planning agency, often with the assistance of consultants and research institutions, reviews and assembles information on the nature and use of the area. This is used to develop reports and summaries for user and public participation, interagency consultation and consultation with appropriate representatives or experts.
2. Public participation or consultation prior to the preparation of a plan. The agency provides information on the nature and purpose of the planning process. It seeks public and user comment on the accuracy and adequacy of review materials and suggestions for the goals and contents of a plan for the area. There should be a parallel program of interagency consultation to establish the views of relevant government departments. Consultants may be used for these tasks but it is important that senior officers of the planning agency are sufficiently closely involved to be able to appreciate and evaluate the positions of the key user groups with interests in the area.

3. Preparation of draft plan. The agency, possibly with the assistance of consultants, prepares a draft plan consistent with the legislation and taking account of information and views obtained in the preceding phase. Specific objectives should be defined for the plan as a whole and for any component zones or subareas within the plan. The agency should also develop explanatory materials for the next phase.

4. Public participation or consultation to review the draft plan. The agency provides information on the results of the initial phase and explains the design and concepts of the draft plan. It also seeks public and user comments on the content of the draft plan. In general the approach and the range of individuals and agencies consulted should be similar to those for the first phase of public participation and consultation.

5. Plan finalization. The agency finalizes the plan taking account of comments received on the draft and submits it for approval and adoption as specified in the legislation.

GOALS AND OBJECTIVES

The key to development of the plan lies in the establishment of goals and objectives to guide the decision-making process. A goal describes a general state or condition which should exist as a consequence of effective implementation of the intended plan. Goals are usually broad statements of intent, for example:

- To preserve an area in its natural state, undisturbed by humans except for the purposes of scientific research
- To protect an area critical to the existence of an endangered species
- To provide a sustainable harvest of fish and other marine resources
- To provide for all reasonable forms of use within an area while minimising conflict between uses
- To retain options for future development of recreation and tourist uses
- To maximize economic return from sustainable use of an area

The process of establishing goals will be a major factor in determining whether an area can be managed as a single entity, or whether a system of zoning should be used. In most situations, but clearly in those that lead to multiple-use planning, there are goals that cannot be applied at the same time to all parts of an area. Zoning, by providing separate areas for different ranges of activities covered by specific sets of goals, generally allows for a wider range of uses. As an example, four immediately incompatible goals which would indicate the need for zoning within a managed area might be:

- Goal 1. Maximise opportunities for local subsistence fishing
- Goal 2. Introduce commercial export fishery
- Goal 3. Preserve coral reef areas undisturbed by humans
- Goal 4. Promote recreation and tourism based upon appreciation of the natural environment

Clearly these goals cannot all be applied at the same site. What is required is objectives that provide guidance as to the extent to which each should apply within a larger area. The objectives should be expressed in quantifiable terms so that the degree to which the plan achieves the objectives should be assessable. Thus, extending the example of the four goals presented above:

1. Maximise opportunities for local subsistence fishing—within 10 km of village settlements
2. Manage commercial export fishery—limit the issue of commercial fishing licenses so that 40% of estimated available catch can be retained for subsistence, local market, and recreational fishing
3. Preserve coral reef areas undisturbed by humans—Preserve 5% of reefs free from human access other than for purposes of approved research or management projects. As far as practicable allocate that 5% of reefs to achieve minimum disruption to other activities and goals
4. Promote recreation and tourism based upon appreciation of the natural environment—In 5% of the managed area, including 10% of the coral reefs of the area, make specific provision for access for tourists and recreational users

The extent to which it is desirable to quantify the objectives in public discussion is a matter of tactics that depends on the nature of the decision making process. Certainly, some quantification is needed within the planning team in order to ensure and assess the balancing of goals in the plan. Ultimately the adoption of goals and their relative weighting through objectives should be a decision of the community reflected in a decision of government. The planning process should be an interaction between government, users, concerned groups, individuals, and technical advisors. The aim is to develop policies for which the relative priorities and the direct and indirect consequences are understood and which are accepted to the greatest extent practicable.

PREMANAGEMENT INFORMATION

The purpose of premanagement information is to establish firmly the case for management and the nature of the issues to be addressed. This involves generating acceptance by the community decision makers of the importance of the structure, processes and amenity of the area and of the need for early limitations on current or potential human activity.

Users, particularly subsistence and commercial users, will often accept the long term logic but employ a variety of strategies to postpone introduction of a management scheme. One of the most common strategies is to require the case for limitation or control to be supported by a body of information much more extensive than that required for initiation or expansion of an activity or an impact. In the area of fisheries, the tragedy of the commons[3] reflects the imposition of so high a standard of proof before acceptance of limitations that economic and biological collapse is virtually inevitable. Kelleher and Lausche[4] state that international experience has shown that it is often a mistake to postpone, by legislation or otherwise, the establishment and implementation of marine management until massive research and survey programs have been completed.

The interaction between the scientist and the manager is critical in the design of premanagement investigations. Time and resources for collection and analysis of information are always limited. Premanagement research and survey objectives should be carefully considered in order to make best use of available research resources. There is always the dilemma that increased information will improve the case for management but at the cost of delays and possible increases in the problems to be managed. Often sufficient information already exists to make strategically sound decisions on the boundaries of areas to be managed, and the degree of protection to be provided within them. Many of the problems, impacts, and solutions are well described in the scientific literature from various sites around the world. Expert literature review can greatly reduce the need for extensive and time-consuming field work in the premanagement phase.

It is not the purpose of this book to establish or recommend particular research or data analysis techniques. Appropriate techniques vary from situation to situation, so, other than the recommendation that careful attention be given to the design of statistically valid survey and data analysis techniques, the discussion in this section concentrates on the questions that need to be addressed in the various stages of establishment and implementation of management.

Typically the physical and biological information is relatively uncontroversial provided statistically valid techniques are used for data collection and analysis. Socioeconomic data is usually more difficult to obtain and interpret. There are two main reasons for this.

The first is that users, particularly those involved in fishing and collecting, are generally reluctant to reveal details of catches and the relative importance of sites. In part this is because they do not want competitors to have the information and in part, in many cases, because they fear that the information will reveal liabilities for tax or government charges.

The second is that marine management is a socioeconomic phenomenon. It arises either to protect resources used for one or more forms of economic benefit, or as a consequence of philosophical, religious, cultural or scientific views of the relationship between humans and the other components of ecosystems. Much of the interpretation of such data as is available depends upon assumptions about the importance of matters such as the value of wilderness or the essentiality of an industrialized economy that may not be shared or even understood by all the groups affected by the plan.

In simple terms premanagement information should briefly address the questions: What is special about the area or resource? Why should we maintain it? What threatens it? What information is needed to make a practical management plan?

The investigations required at the premanagement stage should address three objectives. The first is to establish what is known of the area on the basis of information already available in scientific publications, maps, and government reports. The second is to establish the key issues to be addressed in protecting or managing the resources occurring in that particular marine area. The third is to determine the need for further information to form the basis of plan development, that is, resource, usage and impact information necessary or desirable for design of practical management measures.

Following is a list of questions that should be explored in the existing literature and other information sources on the area to identify critical information gaps and priorities for survey and research.

- What maps, photographs, or satellite images exist of the likely, management area?
- What area needs to be physically surveyed to establish boundaries for a management scheme?
- What is known of tides, tidal, and other currents in the area?
- What water movement information is available to assist understanding of linkages within and beyond the area?
- What is the geological nature of the seabed and coastal land in the area?

- What is the biological nature of the area?
  - -community descriptions
  - -species diversity
  - -dominant biota
  - -commercially important species
  - -endangered or threatened species
  - -changes with time
- What is the cultural/historic significance of the area?
  - -traditional use and impact
  - -traditional understanding
  - -significance of the area to the culture of local or distant human communities
- "stewardship" priorities, for example, caring for the heritage of future generations—recreational significance—aesthetic significance
- What is the international and regional significance of the area?
- What is current use and impact?
  - -subsistence?
  - -traditional?
  - -commercial?
  - -recreational?
  - -research?
  - -tourism?
  - -shipping?
  - -dumping and waste disposal?
  - -alienation, reclamation?
  - -mariculture?
- Who does what?
- What is the socioeconomic importance of current use?
- What changes to use, impact, or environmental condition are likely to occur in the next 5, 10 or 20 years?
- What examples are there of management measures to address similar problems in comparable situations?
  - -limited access?
  - -pollution controls?
  - -impact prediction and control requirements for new or changed activities?
  - -education?
  - -restoration or repair?

Information on key issues builds on initial understanding of the reaction of key groups to the case for management and is likely to concentrate on aspects of human impact upon the structure and processes of the environment, the options for modifying human use and impact, and the feasibility, costs and benefits of the options. Clearly there is some overlap since the background information will identify some if not most of the problems that must be addressed in designing solutions. The design of further information collection for plan development programs involves a tradeoff between the benefits of spending more time in order to improve the confidence of decisions and the costs of delaying management to address identified problems. In the plan development stage the simple question is: How can human use or impact best be controlled in order to protect the area or resource and to maintain or increase its amenity value?

Socioeconomic information is particularly important at this stage but its collection and analysis is complicated since it clearly relates to political and economic strategies that will influence future use and amenity. Key questions are listed below.

Key questions for plan development are:

- Why is the area used?
- What alternatives exist for:
  -food and material for local use?
  -food and materials for sale or trade?
  -cultural recreational activities involving taking or modifying resources?
  -cultural/recreational activities without taking resources?
  -disposal of wastes?
- What groups or communities have interests in the area?
  -traditional users?
  -nontraditional current users?
  -commercial interests?
  -government interests?
  -international interests and obligations?
- What are the apparent impacts of each use type?
- Are there conflicts between those who use or wish to use the area?
- What is the economic basis of current uses?
  -extractive?
  -fishing?
  -collecting?
  -mining?
  -tourist industry?
  -alienation, reclamation?
  -waste disposal?
  -who benefits?
  -who pays?
- What are the economic and social objectives for the area being planned?

-economic growth?
-employment?
-equity: who are the beneficiaries or losers?
-intertemporal equity: will future generations be better or worse off?
-price stability?
-protection of culture?
-regional development?
-self sufficiency?

- Are there any specially difficult or sensitive groups?
  -traditional owners?
  -commercial users?
  -other nations?
  -other national or local government agencies?
  -ethnic or linguistic minority group?

Typically, assessment of information for plan development involves comparison of subsistence, commercial, recreational, and philosophical values. The process is usually complicated by scarcity of comparable information and an understandable tendency for user groups to overstate the value of their own activity in relation to other activities.

For commercial activities there are usually statistics on investment, employment and value of product. These can be used to calculate local and national multiplier effects based on estimated expenditure generated by the supply of goods and services supporting those involved in the commercial activity.

For subsistence activities and for waste disposal, pollution and other activities that may have an impact on the area, but do not depend directly upon its environmental condition, a value may be derived from market costs of preventing the impact or substituting for the goods or services taken from the marine area in question. As an example the costs of installing and operating an alternative system may be calculated for a factory that uses an estuary for the disposal of wastes.

Many recreational activities are relatively difficult to quantify. Values may be derived from the costs of participating in the activity or of taking part in equivalent activities. These values may be further explored by surveys to assess the willingness to pay over and above actual current costs. Such derived values are usually based on assumptions regarding the equivalence of other activities, the effect of other uses on the amenity of the use in question, and philosophical approaches to the importance of recreation and wilderness.

In most situations users and long-time residents are a vital source of information. The fisherman who has, over the years, spent 200 or more days a year catching fish is a source of a vast amount of information on such matters as the distribution of biological communities, the direction of currents, the location of eddies, the nature of the seabed and of change in conditions over time. Such information is often highly selective and rarely quantitative. It is less precise than that collected by a scientist in the course of an expedition or a study of a year or two, but it is no less an important source for plan design. Ideally the two should complement each other.

It is inevitable that a plan will contain elements based on imperfect information. The imperfections often arise from the difficulties of assessing claims and counterclaims by user groups of the importance of areas and activities, and the consequences of limiting or not limiting activities. These will usually be overstated and the decision-making process requires the exercise of judgment in order to develop a "reasonable" solution.

MAKING A PLAN

Most marine areas have historically been subject to plans and policies concerning a number of activities, particularly shipping, fishing and coastal engineering. Usually there is little coordination of such policies although, by default, the sum of policies and plans applying to a marine area constitutes a de facto plan. When the need for marine environment management planning is recognized its scope may fall anywhere in a range from the selection and management of sites for specific purposes such as protected areas or recreation reserves to the creation of integrated multiple-use management regimes over very large areas.

Given the large range of possible contexts, there is no single ideal model for a management plan. The actual form of any such plan depends on the nature of the society and the legislation that authorizes it. The important concept is that there should be site-specific management appropriate to the various parts of the area in question. Where this involves the creation of more than one usage regime or zone, there should be an overall coordinating strategic plan.

Whatever the format, it will be necessary during the course of development of plans to consider available information and to present the plan and associated reports at levels of detail appropriate to address and present issues for the various participants in the decision making process. The checklist below presents headings and brief descriptions of categories of report or information likely to be required in some form in any planning program.

Matters to be covered in planning documents:

- Executive summary—covering the essential issues and necessary decisions. Many of the final decision makers will not have time to read and digest supporting detail.
- Introduction—defining the purpose and scope of the plan and explaining the legislative basis and authority for plan development.
- Definition of the area—a formal statement of the boundaries of the planned area and a geographic description of its setting and accessibility.
- Description of the resources of the area. A summary of information directly relevant to decisions should be included in the report. Detail should be restricted to an appendix or separate document.
- Description of uses of the area. This should concentrate on present uses but should place these in the context of past types and levels of use in the absence of a plan. The description should include social and economic analyses of use.
- Description of the existing legal and management framework such as coastal fisheries, marine transportation and other relevant legal controls on present use of the area. Where they still exist or can be recalled, traditional practices of management, ownership or rights to use marine resources should be described.
- Statement of the principal threats to the conservation, management, and maintenance of the area
- Analysis of constraints and opportunities for activities possible within the area
- Statement of the interagency agreements made, or necessary, for conservation and management of the area. This may usefully include a summary of consultative processes followed in plan development
- Statement of the boundaries, objectives and conditions of use and entry for any component zones of the planned area
- Provision for regulations or decrees required to achieve and implement boundaries and conditions of use and entry
- Provision for regulations or decrees required to establish a permit or licensing system for case by case consideration of significant new or altered activities that may impact upon the managed area
- An assessment of the arrangements, including financial, human and physical resources, required to establish effective management of the plan. This should cover:
  - staffing
  - equipment and facilities
  - training

financial budget
interpretation and education
monitoring and research
restoration and maintenance programs
surveillance
enforcement
evaluation and criteria for review of effectiveness

It is likely that a management plan will be a statutory document supported or supplemented by separate reports, site or activity plans, guidelines or procedure manuals that will be developed during the life of the plan. Later chapters present case details of the development and contents of management plans for the Great Barrier Reef Marine Park. It should be re-emphasised that the form, approach, and contents of management plans will differ in response to site-specific considerations. Examples available in the literature should never be treated as turnkey models, rather they should be analyzed to determine what elements are applicable to another particular situation.

THE FIRST PHASE OF PUBLIC PARTICIPATION AND CONSULTATION

The development, testing and refinement of goals and objectives is a primary function of the first phase of public participation and consultation. On the basis of the initial management information the planning agency, often with the assistance of consultants and research experts, assembles reports and designs other materials for use in the public participation and consultation process.

There are two major purposes for the first phase. One is to present information on the resources and use of the area, on the need for management, and on the objectives and processes of the planning process. The second is to obtain information, to extend and update that presented on resources and uses and to determine whether the goals and objectives are adequate to address community expectations. It is often necessary to develop materials for a number of target groups. It is desirable that each user group becomes aware of the issues and concerns important to the others. A single information package has advantages but because different groups may require information in different forms this is often impractical. The groups to be contacted should be identified early in the planning process. They are likely to include:

- Local resource users
- Researchers and technical specialists
- Industry lobby groups
- Conservationists
- Non-government organisations
- Community opinion leaders
- Local government
- Other government agencies

The nature and strategy of the information and opinion gathering process depend upon the social structure and traditions of the society and communities to be consulted. The design of materials and programs appropriate to the communities to be consulted is a critically important process. Materials must use language, images, and idiom that reflect the relationship between the individual, the community and the environment. The example of the Great Barrier Reef Marine Park, presented in the next section, reflects a widely scattered community that can be reached through electronic and print media supported by follow-up contact. Other situations may involve working with communities where there is a social tradition of face-to-face contact and no widespread distribution of information in written or electronic form. Here the approach of user involvement should be adapted to convey information and seek information and opinion through socially appropriate techniques, which might include drama, song, storytelling, or church meetings.

The programs should be developed with people who know and understand the lives and the values of people in the communities. In an open democracy it may be reasonable to expect that individuals will feel free to comment to government officials or consultants. Nevertheless, it can be the case that elected officials or self-appointed community representatives will convey information and views that do not take account of majority interests. In such situations there may be strong distrust and cynicism regarding the process and extreme reluctance to convey any information or opinion. In some situations, and particularly in rural or fishing communities little affected by economic development, there is a strong hierarchical structure with hereditary or appointed leaders who speak for the community. Here, approaches to other community members for additional information and opinion will generally fail and, as a breach of accepted protocol, they will cause embarrassment and distrust.

The results of each response to the public participation program can be analyzed in terms of the interests covered and the specific comments on particular aspects of the plan. There is a range of options in the design and conduct of the process from a formal referendum of all residents of the area, through a statistically designed randomized opinion sampling program to consultation of community leaders and known key users or opinion leaders to an optional approach in which the decision to participate is a matter for the user. The selection of an option or combination of options should be based on sound understanding on the structure and decision-making process of the communities involved. Whatever the selection, it will clearly affect the way in which the information can be analyzed and used.

A referendum will determine the public view of the answers to a small number of specific questions. It is vulnerable to problems of question design. Further, unless the issues of marine environment management are uppermost in community thinking, the signal of the response of people to whom the issues are critical may be swamped by the noise of those to whom they are unimportant or poorly understood. In a party political framework, the central issue of a poorly understood or noncritical referendum may be submerged by political point scoring.

Properly designed and conducted opinion sampling will provide a reasonable analysis of the views of the community at large and of the sectors of the community most affected by the marine environment management issue. Unless a substantial proportion of the population uses or appreciates the issues of the marine environment it is likely that stratified sampling will be needed in order to identify and sample an adequate representation of the views of key minority groups.

Well-publicized optional public participation has the advantage that concerned individuals or groups are motivated to become involved in the planning process. The sample is not statistically designed and there is no means to prevent an individual providing multiple responses under different names or a group organizing a mass response to support a particular position. The results should be treated with some caution to avoid the assumption that the relative proportion of responses advocating a particular position is necessarily representative of the strength or distribution of that position in the community or among those concerned with the marine environment. The advantage of the technique compared to sampling is that knowledgeable and concerned individuals can contribute to the planning process and that in so doing they may identify with the objectives and form of the eventual plan.

Consultation of key individuals and community leaders is an important part of the political planning process, which will generally produce a good understanding of the issues and of the views of government and administration. Its value in the technical process of gathering detailed information from users depends on the socioeconomic importance of the marine environment to the influential sector of the society. In technologically advanced societies the business and social groups that tend to become involved in politics and administration are generally not involved with the marine environment other than in a recreational capacity.

In practice a combination of the techniques is probably needed in order to achieve an understanding of the importance of a marine area to the communities that use it, and of the views of people more distant from the area but concerned with its management.

There are several component purposes that should influence the design of the materials and program for initial public participation and consultation. These include:

- Presenting information concerning the issues, benefits of management and costs of failure to manage
- Presenting information summaries prepared by the planning agency concerning the nature and use of the resources of the area
- Encouraging members of the community to discuss the issues and develop views on desirable future use and management
- Seeking information on the adequacy and accuracy of the information and understanding indicated in the summaries provided by the planning agency
- Seeking information on matters and approaches which the community considers should be incorporated in a plan for the area

The design of the materials to present new and often complex information and concepts in a manner acceptable to and comprehensible by users who may have operated for years or generations in the area frequently involves a major educational task. Fishermen and mariners, used to dynamic changes in the state of the sea surface and in the availability of fish or other resources, may find it difficult to accept the concept that human actions can influence the nature of the biological communities of coastal waters and the open ocean. The concept that there are limits to the resources of the sea is a marked departure from the fundamental beliefs and traditions of many seafaring communities.

The design of materials and questions to obtain information without biasing the result is a similarly complex task. There is an extensive literature on survey and questionaire design.[5] For remote communities which may not be exposed to major economic activity, and that often have different cultural and linguistic traditions to those who are seeking information, the task is particularly complex. The fundamental questions may be clear. The appropriate method of asking them in order to obtain valid information involves substantial understanding of the community, its cultural traditions and use of language. Without this, the result may be a reflection either of an intent to present what the questionee believes the questioner wants to hear or, worse, of a wish to mislead the questioner.

The equally important design task is to arrange a program to ensure that the materials reach the intended targets and to establish a timetable for consultations, public meetings and other activities. The program should be designed to ensure that no individual who has interests in the area to be managed can escape being aware of the planning process and of opportunities to learn more of, and to make comment on and contributions to, that planning process.

PREPARING A DRAFT PLAN

The aim should be to make the plan as simple as practicable consistent with providing the range of degrees of protection considered necessary and avoiding unnecessary restrictions on human activitiy.

The starting point is to develop an understanding of the constraints and opportunities for the range of possible activities within the boundaries of the area. Some activities can take place anywhere, others only at specific sites within the area. Several potential activities are incompatible with each other. Constraint analysis is essentially a matter of creating a matrix of the compatibility of each use against other uses and the desired condition or setting. For each entry point it should be possible to determine whether there is no conflict, possible conflict, probable conflict or absolute incompatibility. This should thus help to identify the extent that activities can coexist and thus the number of zones or categories that may be needed. Following is a list of activities that might be considered in a constraint analysis for a coastal area:

Preservation of undisturbed natural environment
Protection of breeding area of endangered species
Protection of breeding area of commercially important species
Scientific research
Nonextractive recreation
Nonextractive tourism
Extractive recreation (fishing and collecting)
Extractive tourism (fishing and collecting)
Subsistence fishing
Commercial fishing
Mariculture
Commercial collecting (e.g coral, shells, aquarium fish)
Shipping
Extraction of construction materials (rock, gravel, sand)
Dredging of navigation channels
Construction of tourist facilities

Construction of navigational aids
Residential construction
Light industry
Heavy industry
Coastal mining
Seabed mining
Hydrocarbon drilling
Stormwater runoff disposal
Domestic sewage disposal
Municipal sewage disposal
Industrial waste disposal
Solid waste dumping
Alienation, reclamation, marina, harbor or breakwater construction

The identification of specific opportunities is a geographic information task. It may be done by map overlays or through a computer-based Geographic Information System, or GIS.

The planning team should build from the basis of knowledge of the resources, and of constraints and opportunities for the range of possible uses, to design appropriate management controls. Frequently this will involve zones or different controls in subsets of the management area. A specific statement of purpose should be developed for each zone.

It is desirable that the planning team should develop and work to a series of guidelines. These should generally be expressed with the preamble "as far as practicable." They should express a contemporary interpretation of "reasonable use" that should ensure that the range of accepted uses is properly considered and provided for. These guidelines taken together should thus cover all the uses and objectives to be provided for in the plan.

Guidelines will often be in mutual conflict. Thus two guidelines may be "as far as practicable, areas close to coastal towns should be available for recreational fishing" and "as far as practicable an area close to each coastal town should be set aside for activities other than fishing and collecting." In such cases the balance of practicability will be assessed on the basis of initial information and the results of the first phase of public participation and consultation. It will be tested in the second phase.

The plan thus prepared should be consistent with the requirements of the legislation. It should present a solution that the planning authority considers workable and acceptable. The desirability and acceptability of the plan will be tested in the second phase of public participation and consultation. It can be expected that the affected groups will react and that if the eventual result produces no change in response to their reaction they will question the value of the consultative process.

Depending on the decision-making process of the society concerned, it may be tactically desirable for the draft plan to err slightly in the direction of controls and restrictions in the expectation that reductions will be argued by user groups. This can provide an opportunity to try solutions that test the relative concerns and priorities of users in the face of slight overrestriction. For example, the draft plan may include one or two or more alternative sites for a particular restrictive use than ultimately thought necessary with the expectation of identifying and removing the least appropriate through public reaction. The margin of draft design overcontrol requires careful judgment. If it is too great, the affected user group may reject the participatory process concluding that no heed was paid to information and proposals in the initial phase of public participation.

THE SECOND PHASE OF PUBLIC PARTICIPATION

The process is similar to that for the initial phase. The purpose of the program is more easily explained to the public. Where the first phase explained the concept and sought information and management suggestions, the second seeks reactions to a plan containing specific proposals. Individuals and interest groups can study those proposals and consider the degree to which they address their concerns.

As with the initial phase, materials should be designed to achieve simple and effective communication with the range of user and interest groups. Similarly, a timetable of meetings and consultations and other activities should be established as a framework for public discussions on the draft plan.

The component purposes that should influence the design of the materials and program for public participation and consultation to review a draft plan include:

- Summarizing the results of the first phase of public participation with particular emphasis on presenting the expressed views and priorities of the various user or interest groups
- Presenting updated information reflecting new knowledge of resources and usage resulting from or obtained since the first phase
- Explaining the goal and objectives of management
- Listing the guidelines used in plan design.
- Stating briefly the reasons for the selection of areas with substantial restrictions of activity
- Encouraging members of the community to discuss the issues and present views on the adequacy of the goals and objectives of management and of the plan in meeting those goals

- Seeking specific reaction to alternative solutions presented in the plan or accompanying materials
- Seeking alternative solutions that are consistent with the goals and objectives but that have less impact on the amenity of users

The processes of analysis and reporting will be similar to those for the initial phase although the task may differ slightly since participants are reacting to specific proposals rather than providing information and general suggestions. The strength and degree of acceptance of views or reactions to specific aspects of the draft plan may be critically important factors in decisions on changes to be made in finalizing the plan. The analysis should reflect the strengths and limitations of the information gathering technique in terms of representation of views of the community or specific sectors of the community.

PLAN FINALIZATION

Analysis of the public review of the proposed plan will reveal the degree of general acceptability of the plan, which should be reasonably high if the plan reflected the views and information collected in the initial phase of public participation. It will usually reveal some generic dissatisfaction from one or more groups that consider themselves disadvantaged or ignored in the draft. It will also shed light on specific situations where the solution proposed in the draft plan should be revised. Such situations arise where new information emerges because the proposed solution would prevent a use or activity that was not previously declared or identified. They also arise where the draft plan contained a marginal decision or an optional choice of site for particular activities.

Revision of the draft plan to produce a final version for official adoption is again a matter for some judgment in the light of community attitudes and expectations of participatory decision making. If there is no change between the draft and the final the community is likely to view the consultative process as insincere. Conversely if there are numerous changes there may be demand for a further phase of consultation because the final plan does not resemble that which was commented upon. Changes between draft and final plan will generally address problems identified in the draft by one user or interest group, but, unless care is taken, they may create problems for another by changing a situation that was seen as satisfactory. Some informal consultation with representatives of key user groups may be needed in order to ensure that the solution to one problem does not result in the creation of a greater problem.

The formal adoption of the measures to be incorporated in the final plan is usually a process covered by legal and constitutional procedures. The result is usually one or more documents written in formal legal language that describe the boundaries of the managed area, and any component subareas within it, and set out the regulations required to give effect to the conditions of use or entry developed in the plan.

It is often the case that the implementation of the plan will be the task of a different group of people from that which developed it. The first tasks of management will generally involve the development of materials for staff training and public education on the provisions of the new plan. It is therefore important that in addition to the formal finalized plan, the planners should provide the management team with reports to explain the plan, the key decisions and the major outcomes of the public participation processes.

Chapter 7

Implementation Of Management

The logistics of management should be a factor in plan design. Nevertheless implementation will involve a phase of detailed preparation. The preliminary phase, which should take place while the management plan is being finalized, involves the development of detailed programs for recruitment and training of staff, for acquisition of equipment and management facilities, and for establishing ongoing programs of day-to-day management.

The overall management program will generally have seven key elements:

- Training—to ensure that those who implement the plan understand it, their duties under it and the management approach they should use
- Education—to ensure that the public and users of the managed area are aware of the existence of the plan and their rights and responsibilities under it
- Surveillance or activity monitoring—to assess how people are using the area and the extent of compliance with the plan
- Enforcement—to follow up infringements of the plan and by education or legal action seek to ensure that they are not repeated
- Monitoring—to provide up-to-date awareness of the condition of the managed area and of the impacts upon it from human use and other factors
- Impact prediction and management—to assess in advance the likely impact of significant new or altered uses proposed for the area, to determine whether such uses are consistent with management objectives for the area, and if they are so, to establish conditions under which such uses may be allowed
- Review—to revise and improve the plan, after it has been in operation for a defined period, in order to take into account the results of surveillance, monitoring, increased understanding of the nature of

the area, and the impact of use and of changes in public attitude to, and perception of, the area

TRAINING

The initial implementation of marine environment management usually involves establishing a special management unit. This may be done by recruitment of new staff, but usually it involves at least some transfers from units with parallel skills, such as national parks, fisheries or coastguard units, which are likely to have objectives that differ to some extent from those of the new management plan. Training and workshops to develop a consistency of objectives and approach in the initial management staff are particularly important because that staff sets precedents of attitude and user relationships that may be difficult to alter. If the initial training is successful it will provide the basis for induction training following recruitment of subsequent staff specialists and volunteers. Ongoing training continues to be important so that staff update their skills and maintain their interest and commitment to marine management.

The needs and the appropriate methods of providing information differ between the levels in the management network.[1] For each level, training should draw upon a body of common information to establish the setting for management responsibilities. Against this background it should address the particular knowledge and skills required to carry out those tasks. The common information pool will include:

- History—why management was introduced
- Description of the managed area—why the area is special
- Objectives of the management plan
- Relations with users and other agencies with interests in the area

Particular knowledge and skills will range from use and maintenance of vessels and communication equipment for field staff to relevant international conventions and constitutional issues for senior managers. General background information for management staff covers a working familiarity with a range of disciplines and more specific expertise in one of them. At the start a few, generally senior, staff may have specifically relevant degree, diploma or certificate qualifications. Nevertheless, most training, particularly in the early stages of establishing marine management programs, is a matter of short courses or work experience to build upon the basis of earlier secondary or tertiary training.

For many developing nations the problems of providing training, particularly for training field and junior management staff, are made greater by language and cultural factors. The options are to train in a language other than the trainees' primary language, which is inefficient, or to develop and translate specific training materials, which is expensive.

Different forms of training are needed for different levels within the management structure. Some may be recognized as liaison and intragovernment relations more than training but since the objective is to achieve the understanding needed for effective implementation of management they are as much training as that provided to a patrol boat crew. The categories for specific training include:

- Top policymakers, heads of state, government ministers, parliamentarians, congressmen, local government and community leaders: for understanding of the purpose of management and of the importance of the managed environment
- Top management within the agency: for detailed understanding and philosophical framework of the management task
- Top management in related or interacting agencies: for understanding of the interaction of management with the function or their own agencies
- Divisional heads within the agency: for understanding of the roles of management and of the interactions of the components of the management task
- Management staff: for understanding of the roles of management and the skills to undertake and supervise the conduct of management tasks
- Rangers, Patrol Boat Crew: for understanding of management and for the development and reinforcement of special skills for management tasks
- Volunteers: for understanding of management and for the development of special skills for the volunteers program

The range of options for presentation and development of knowledge and skills for management includes:

- Induction briefing—a half day presentation on the goals, objectives, organization and general operational basis of management. This may be modified depending on the role of the inductee in the system. It is often the only opportunity for providing essential basic information to ministers and other top policymakers
- Familiarization tour—a specific program providing for inspection and briefing on the functions and procedures of the various parts of the

organization from policy development and planning to field operations. Generally such tours last a week or two. In house they serve an important introductory training function. They are also a good means of providing and obtaining detailed information in the context of international aid, training, and collaboration programs

- Seminars, workshops or short courses—the distinction between these terms is largely a matter of personal or institutional usage. The form and function can cover a wide range. At one extreme, for example, remote sensing or a similar topic requiring access to sophisticated equipment, they will involve a program at a specialized training location. At the other they may involve numerous site inspections and be conducted on a traveling basis

  - Seminars may be structured specifically for training, in which case they involve a specialist who presents material on a particular issue or problem solving technique and promotes discussion on the relevance and applicability of the information to situations faced by the trainees. Often they take the form of professional updating where a researcher or management expert presents current or recently published technical information. Typically they are of short duration, an hour to half a day.
  - Workshops may be presented as training projects where a trainer takes trainees through set exercises designed to develop functional skills in applying a particular technique or using complex equipment. Alternatively they may be structured as a collective problem solving activity where an organization or a group of individuals with similar functions in different organizations tackle an issue for which a policy and practice have to be developed. In this case they should develop a solution and provide the basis for training on the new policy. Typically they occupy a few days.
  - Short courses are led by one or more specialists who take participants through a program of lectures and activities designed to introduce or refine the knowledge and skills needed to address a particular problem or field of activity. Depending upon the topic, the duration is very variable in the range from a few days to a month.

- Secondment—in this case an expert with well developed skills from another agency works for a period within the trainee agency. Structured deliberately, for example, by preparing and assigning specific staff to work with the seconded expert, this can be an effective method of transferring skills.
- Attachment—the inverse of secondment. In this case trainee staff

work within an expert agency for a period. They can thus become familiar with the procedures and equipment used by that agency and should, on return to their own agency, be able to apply relevant parts of that experience to their own situation

In the longer term it can be expected that more centers specializing in research and specific tertiary training for coastal and marine management will be established. These will not remove the need for shorter in-house or secondment or attachment training. They should however lead to better documentation of management issues and of successful approaches to management problem solving and thus speed the processes by which experience is shared and expertise developed and applied in management agencies.

A recent initiative to address the problem of syllabus material for translation for own-language short courses and to provide rapid information sharing was undertaken by the Coordinating Body on the Seas of East Asia (COBSEA), within the UNEP Regional Seas Program. This involved the development of a series of modules of basic and applied training materials for South East Asian Marine Parks staff. Based upon Australian materials, previously prepared draft English language modules were further developed at a workshop of senior staff of Southeast Asian Marine Parks. The materials, as yet unpublished, were designed for subsequent translation to own-language and local example versions for use in a range of training settings for staff at two levels, middle management officers and rangers or field staff.

USER AND PUBLIC EDUCATION

Hudson produced a synthesis of practical approaches to user and public education for marine environment management.[2] That material is reproduced here with some development to broaden it from the focus on coral reefs.[3]

People cannot be expected to comply with a management plan unless they know that it exists, are aware of the effect that it should have on their activities, and believe either that it will benefit them in some way or that the costs of infringing the plan will exceed the likely benefits. Education should seek to achieve the greatest possible user cooperation and support for management and its objectives.

In a typical coastal or shallow marine management situation there is a wide range of users and interest groups whose perceptions of the need and priorities for management differ considerably. Each such interest group is likely to need a specific approach. The first step is to identify target groups and key messages. These should emerge during the processes of public participation and consultation that lead to the development of a management plan.

The target groups may include:

- People in geographic units—villages, provinces, states
- Specific user groups—commercial and recreational fishers, tourist operators, scientists, conservationists
- Developers, engineers and planners—industry and government
- Decision makers, policy advisors and staff of government agencies
- International agencies
- Neighboring states or nations
- Special subgroups—e.g. linguistic minorities
- Educators—from primary to tertiary levels

The specific approach for a target group follows from determining its concerns, level of common interest and understanding of aspects of use and sustainable management of the environment. This provides the basis for developing appropriate messages as illustrated in Table 7.1.

Methods for the delivery of messages may cover a wide range from formal messages from community leaders, through current interest stories in print and electronic media, paid advertisements, education in institutional programs, local drama and story telling, to consciousness raising through T-shirts, badges or stickers.

Table 7.1
An example of some Target Groups and Message Contents concerning Management of a Coral Reef Area

Target Group	Message Content
General Public	Nature of coral reef environment; need to protect reef areas
Local fishers	Economic benefits of management specific provisions of plan related to fishing
Tourist operators	Reef protection conserves the natural asset for tourism; specific provisions of plan relating to tourism
Other government agencies	How the plan interacts with their responsibilities

Education programs are generally very costly in terms of time and financial resources. It is important to assess the effectiveness of the methods used by determining changes in knowledge and attitude attributable to the program. A critical test is the level of adherence to the management plan and support for the concept of management.

SURVEILLANCE OR ACTIVITY MONITORING

Surveillance is a term that has sinister connotations for some, but it is a necessary subset of monitoring. It is used here to distinguish those activities that record what people do to the managed marine system from other monitoring that records how the system changes in response to the combination of human activity and underlying physical and ecological dynamics.

Because the purpose of a management plan is to control the ways in which humans use, or have impact upon, a marine area it is critically important to monitor the level of human use of the area and of activities outside the area that have impacts within it.

Surveillance should result in records of use of the managed area and of apparent infringements of the management plan. These records provide a basis for enforcement operations and enable the effectiveness of the plan and its implementation to be assessed.

In a small management area the manager may be able to see the entire area and to observe all visitors and uses. More usually surveillance requires patrols in boats and, in large areas, in aircraft. Such active surveillance techniques are expensive.

Alternative techniques include remote and passive surveillance. Remote techniques include photography, radar, infrared photography or images for night observations, hydrophones and aircraft or satellite electronic remote sensing. Passive surveillance techniques include user log books, permits or licenses issued by management and reporting by volunteer observers or rangers. Remote and passive techniques can be cheaper and more effective than direct observation but clearly there is a need for ground truth checking to assess the reliability of the information collected.

Surveillance patrols serve an additional purpose to data collection. They are often the most obvious sign to users that management is occurring and that plan infringements will be recorded. This is an important management factor since supporters of management are likely to lose their enthusiasm if the impression develops that infringements are undetected and unpunished since such supporters see themselves disadvantaged by their adherence to management provisions while the less scrupulous profit from illegal activities such as fishing in protected areas.

ENFORCEMENT

Effective enforcement is essential if a management plan is to succeed. Recorded infringements should be followed up by advice and counseling or by prosecution. The choice depends on a number of factors that are largely local.

The advantage of the option of advice and counseling, particularly for first offenses in the early years of existence of a plan are:

- Maintenance of good relations between management and users; The infringement may have occurred because the offender was unaware of the plan or its precise provisions. Prosecution under these circumstances may be seen as unreasonable and raise unnecessary hostility to management
- Effective use of staff and resources; Preparation of documents for prosecution can be a complex and expensive operation. A prosecution that fails because of a technical error can reflect poorly on management
- Effective action can be taken by counseling to prevent recurrent infringement. This is particularly true where the assembly of complete evidence to the standard necessary for prosecution has not been possible

In many cases a large measure of compliance can be achieved by peer group policing whereby users who are aware and supportive of the plan counsel and encourage others to comply.

There is always a need for some firm management action to prosecute and punish gross violations and repeated violations. Such action is important if management is to keep faith with law-abiding users whose actions support the plan.

Enforcement staff need an aptitude for detail and procedure strengthened by special training to ensure that they collect and record evidence in a manner acceptable to the judicial system used for prosecution. Enforcement staff are thus often specialists, but they also play an important role in surveillance and the collection of information on the attitudes of users toward the management system.

MONITORING

Monitoring involves the collection of information to record the condition of the managed area and the effect of management upon its natural resources. Data collection for monitoring can lead to large and costly programs. It is essential that monitoring be carefully planned and that the techniques adopted are statistically sound. Strategy and sample design for marine ecosystem monitoring constitute a specialized field. It is often the case that monitoring, although part of management, has such specific requirements in terms of training, equipment and sampling schedules that it is best conducted by a specialist subcell of the management unit or under contract by an external specialist agency. Whether conducted by an internal specialist unit or by a contracting laboratory, it is important that senior staff of the managing agency includes at least one person with sufficient training to appreciate the sampling issues and supervise the program.

An essential difficulty is that the natural system is dynamic. Severe storms, exceptional freshwater runoff, intrusion of different water masses as a consequence of current variations and major temperature variations occur in the context of natural fluctuations in the physical interaction of atmospheric and oceanographic forces. Changes may be measured, but the experimental design should address the issue of determining human-induced change against the background of natural dynamic change.

Some basic guidelines can be applied:

- Monitoring should focus on factors that are likely to be particularly responsive to changed conditions
  - occurrence or abundance of key species
  - changes in reproductive and recruitment patterns.
  - changes in behavior
- The factors to be monitored should be selected so that the success of management in achieving defined objectives can be tested. Thus:
  - If the objective is to maintain sustainable harvest of fish, monitor fish catch, fishing effort, target species population, average fish size, and other species populations at key sites
  - If the objective is to maintain or restore sea-bed communities, monitor cover and species diversity of benthic species at key sites
  - As far as possible factors or species monitored should be distinctive and easily recognized.
- Time series data is essential for monitoring. Data collection techniques should be standardized. Changes in technique should be introduced only for clear methodological advantage, and their introduction should be done in such a manner that the relationship of the time series of old and new techniques can be established
- It is generally appropriate to establish some key or detailed monitoring sites that may be the focus of major research by specialists. More general cost-effective coverage may be provided by techniques that enable non-specialist but trained field staff to collect data and infor-

mation that can be evaluated by specialists against the background understanding of the key sites

IMPACT PREDICTION AND MANAGEMENT

Plans are usually developed on the basis of the status quo whereby existing uses, unless specifically forbidden or restricted, are likely to continue and to expand with increases in human population or access to the area. Planning will usually identify some major activities that could potentially take place in the area in a range of forms or scales with impacts ranging from trivial to gross. The control of such activities in order to avoid gross or otherwise unacceptable impacts may be provided through a permit system.

It is also important that a management or zoning plan can make provision for prior evaluation of changed or new forms of activity that are likely to be of a nature, extent or potential impact that could not be anticipated at the time of plan preparation.

In both situations, permitting can provide for prior evaluation, on a case by case basis, of each proposal to conduct such an activity, in order to decide whether and under what conditions it should occur. Such a decision should follow a process that identifies the impacts that might occur, quantifies them and then assesses whether, on balance, the benefits outweigh the disbenefits. Claridge[4] notes that, in practice, given the constraints of lack of time, financial resources, and technical expertise, impact assessment is often reduced to determining whether a proposed activity will be likely to produce any unacceptable impacts and particularly whether the activity will be sustainable or will destroy the resource upon which it or other activities depend.

Impact assessment has developed over the past three decades from an initial "once and for all" process. This sought to identify and predict the extent of all potential impacts with a high degree of confidence and to identify necessary controls to prevent or limit specific impacts. Today permitting processes for major projects are likely to distinguish between the one-off impacts of construction or establishment of facilities and the ongoing operational impacts. In both cases they may establish a management regime of adaptive environmental assessment with continuing monitoring and resultant ongoing adjustment of management conditions as indicated by the results of monitoring.

The processes of impact assessment should take into account the likely impacts on the structure and processes of the environment, and the interaction of those impacts with the amenity and impacts of other existing or potential uses.

Claridge provided a checklist for the consideration of reef related development proposals that is applicable in other marine environments.[5] His list is based on paragraph 4.1 of the U.S.Environment Protection Administrative Procedures.

1. Summary

    - Title of project
    - Name of proponent
    - Project objectives and background
    - Environmental impacts
    - Possible safeguards, alternatives and monitoring procedures
    - Explanation of contents of Environment Impact Statement

2. Introduction

    - Scope of study
    - History and objectives of proposal
    - Justification for the proposal
    - Prudent and feasible alternatives
    - Status, consents required/received
    - Status studies required/undertaken

3. Detailed description of proposal

    - Maps, diagrams, photographs as appropriate
    - Location and proposed tenure
    - Duration of site or activity development
    - Expected life of project
    - Experience and credentials of operator
    - On-shore support facilities
    - Vessel or structure design
    - Mooring of structural foundations
    - Marina or smallcraft moorings
    - Recreation opportunities for staff or clients
    - Servicing arrangements
    - Maintenance practices, hygiene, and antifouling
    - Effluents and emissions
    - Plans for expansion or further facilities
    - Plans for removal and making good at end of useful life

4. Economic issues

    - Market demands
    - Service requirements from regional economy
    - Employment opportunities

5. Description of the environment

    - Physical structure
    - Climate
    - Currents and tides
    - Biological communities
    - Local, regional or globally significant elements

6. Current activities or impacts at site

    - Existing commercial and recreational use
    - Current demands and likely future demands for use and access

7. Regulations governing the development

    - Tenure
    - Zoning
    - Applicable design, operation and emmission standards
    - Consultation with local, regional or national government
    - Applicable international conventions

8. Environmental impacts during establishment or construction

    - Disturbance of seabed and sediments
    - Disturbance of marine and other life
    - Noise and vibration levels
    - Effect of accommodation for construction crews
    - Volume of construction traffic, vessels, aircraft
    - Monitoring plan

9. Environmental impacts during operation

    - Proposed activities
    - Chemical impacts on water quality
    - Impacts on turbidity
    - Impacts on water temperature
    - Changes to water movements

- Arrangements for disposal of wastes
- Content and method of disposal of effluents
- Antifouling measures
- Provision for technical and human error accidents
- Provision for extreme weather events
- Primary biological impacts of activity
- Incidental biological impacts, e.g., of noise, vibration, emissions and shading
- Operational, impacts on amenity of other users
- Impacts on adjacent areas
- Shore-based social and economic impacts
- Demand for shore-based amenities, e.g., sewage or waste disposal
- Energy use and conservation
- Impacts of alternatives

10. Operational safeguards of the development

- Construction and design criteria for on-site facilities
- Control of accidental spills or leakage of harmful substances
- Control of recreational activities
- Operational procedures for heavy weather
- Contingency plans for safety of staff or clients
- Evacuation plan and decision criteria for evacuation
- Monitoring plan

11. Consultation and information sources

- Details of consultation and comment from local community and other users of the area
- Description of relevant studies completed or under way
- Source of information
- Consultation with government agencies

For most development or activity proposals some or many of the categories of information indicated by the checklist will not be relevant. Nevertheless the checklist represents an attempt to anticipate all the types of impact in a marine environment. An initial judgement can then be made on the possible forms of structural, process or amenity impact and a list can then be made of relevant information requirements for detailed assessment of the development.

REVIEW

Once a plan has been in effect for a reasonable period of time, say three to seven years, it is appropriate to review the plan and its implementation in order to determine whether the objectives of management are being achieved and to decide what changes if any are needed.

The primary sources of information for plan review are likely to be surveillance and monitoring, although new information from local or international scientific or socioeconomic research may be relevant. The interaction of surveillance and monitoring should provide an assessment of the effectiveness of the management plan. This assessment should address effectiveness in terms of achieving the stated objectives for which the plan was prepared. It may also address effectiveness in terms of the plan providing for changed circumstances and attitudes toward reasonable use since the plan was prepared.

In simple terms, monitoring should determine whether the condition of the managed environment is satisfactory; surveillance should determine the extent to which people are adhering to the conditions set down in the plan. There are then four possible outcomes:

1. The condition of the environment is satisfactory and the plan is being obeyed. This may be interpreted as an endorsement of existing management. The review should focus on likely changes in use and attitude in the period before the next review

2. The condition of the environment is satisfactory but the plan is not being obeyed. Implementation is ineffective and, at existing levels of use, the plan is not addressing issues which affect environmental condition. The review should focus on the necessity for a form of plan to maintain environmental quality in the face of any likely expansion of threats or developments in the next few years

3. Monitoring indicates environmental deterioration although the plan is being obeyed. This indicates that the plan does not adequately address the factors responsible for environmental deterioration. Either, the plan has not effectively addressed an issue within its jurisdiction, or the causative factor is beyond the scope of the plan. The latter case would apply if the causative factor were pollution from a source beyond the jurisdiction of the planning or management agencies. The review should explore the causative factor. If the cause can be identified as lying within planning jurisdiction, the review should focus on the development of new measures to be incorporated into a revised plan. If the cause lies beyond existing planning jurisdiction, the review should focus on legislative or diplomatic measures necessary to address the issues

4. Monitoring indicates environmental deterioration and surveillance shows that the plan is not being obeyed. The failure indicated by this result may arise from inadequacy of the plan, inadequacy of management, or a combination of both. Lack of compliance with the plan may result from inadequate public education, so that users are not aware of the plan; inadequate enforcement, so that those who are inclined to infringe the plan understand that they can do so without risk; or unnacceptability of the plan, so that the community as a whole sees no merit in it and no need to support it. In any review it is necessary to identify the apparent causes of failure and thus to determine the extent to which revision of the management plan may address the problem.

The process for plan review should be similar to that for initial plan development.

CONCLUSION

The process of initial plan development is usually the easiest step in environmental management. As an innovation and as a measure to address a perceived problem, it is likely to attract the support of politicians and concerned sectors of the public. The process of establishment of the initial plan may be costly but it is a one-off project, usually with a definable conclusion.

Implementation of management plans is more difficult to achieve and sustain. It is an ongoing process with no obvious conclusion, requiring recurrent funding from government or from the user community, through some form of levy or fundraising. The plan may attract supporters by providing or maintaining opportunities for activities or amenity regarded as important by some sectors of the public. It will almost certainly attract opponents when its enforcement prevents groups or individuals from carrying out intended activities. If it is effective marine environment management should lead to changes in attitudes toward the environment and the appropriate balance of uses. It should then adapt to reflect changed attitudes and values.

The excitement of establishment must be followed by sustainable management with procedures that ensure fairness and consistency of decisions. Like other ongoing procedures marine resource management involves the accumulation of precedents and case histories that interpret details of the underlying strategic plan. The review of management plans must be accompanied by regular review of management procedures and policies to ensure that the two remain consistent with each other and that the whole mechanism of management is responsive to change.

Chapter 8

The Great Barrier Reef: A Brief Introduction

GEOGRAPHY

The Great Barrier Reef is the name given to the system of coral reefs and related environments that occurs on the continental shelf of Northeast Australia from just south of the Tropic of Capricorn northwards to the Torres Strait and Papua. There is no clear northern end to the system because as it approaches Papua it merges with the coastal and island reef systems of Papua and the New Hebrides. These, in turn, extend northwards without clear breaks toward the Philippines. Hopley[1] adopted the generally followed convention of regarding the Murray Islands Group in the Torres Strait as the northern limit of the Great Barrier Reef.

Maxwell[2] identified three major regions within the Great Barrier Reef province. The Southern (Fig 8.2) extends from the Swains group where the continental shelf is broad, to 150 nautical miles from the mainland coast northward to approximately latitude 21° 30' S where the continental shelf is 70 nautical miles wide.

Tidal ranges are high and there is a broad coastal lagoon 50–60 meters deep. This too gradually narrows from 70 nautical miles in the south to 30 nautical miles in the north. The coastal lagoon separates the offshore areas with consistently high salinity, low nutrient, clear, oceanic waters with extensive reef growth in shallow sites from the inshore areas. The closest major reef growth occurs between 10 and 50 nautical miles from the coast. The inshore waters have variable salinity depending on levels of rainfall and coastal runoff. They are frequently turbid with sediments brought by the river systems and resuspended by wave action generated by the southeast trade winds. The inshore waters consequently support relatively limited reef growth, mainly in the form of fringing reefs on islands or rocks of continen-

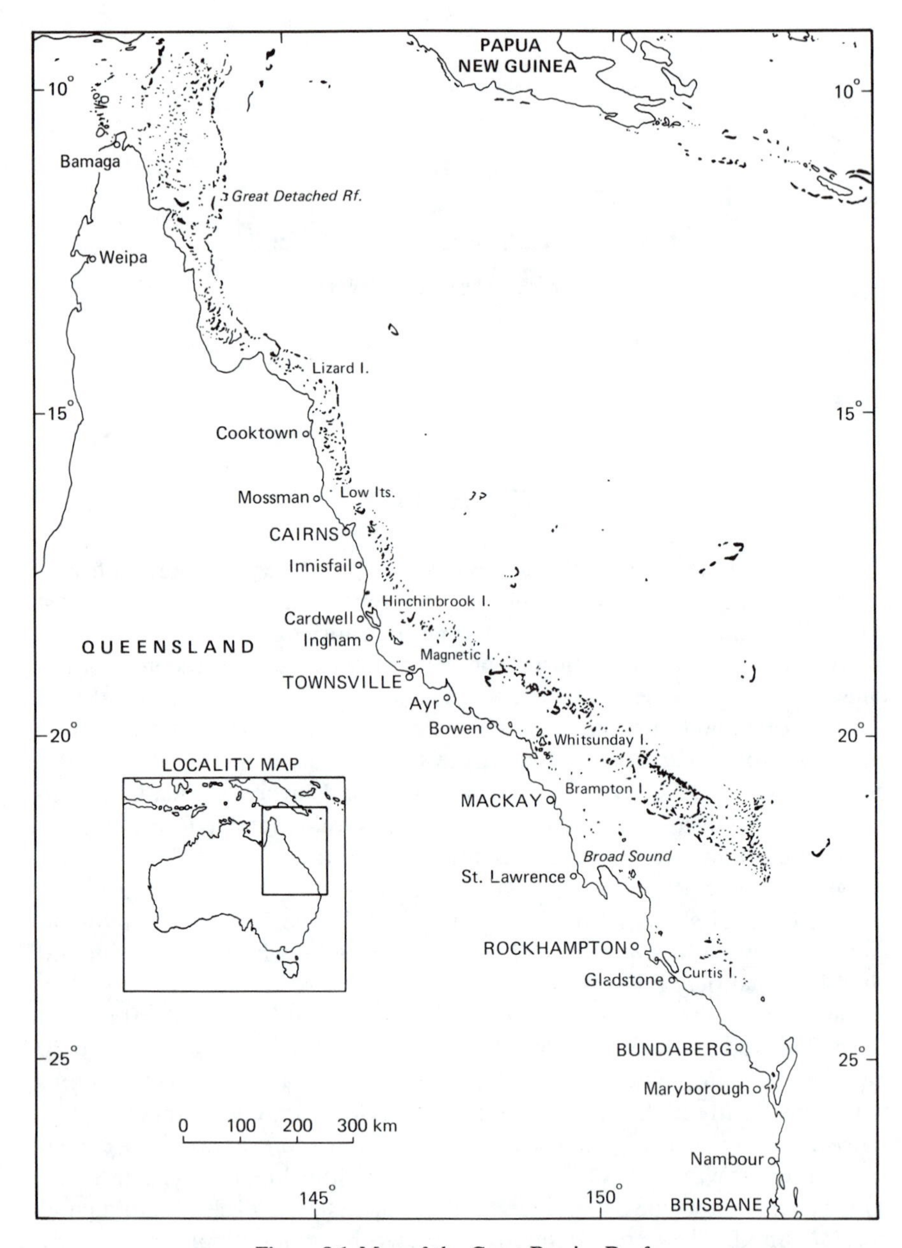

Figure 8.1 Map of the Great Barrier Reef

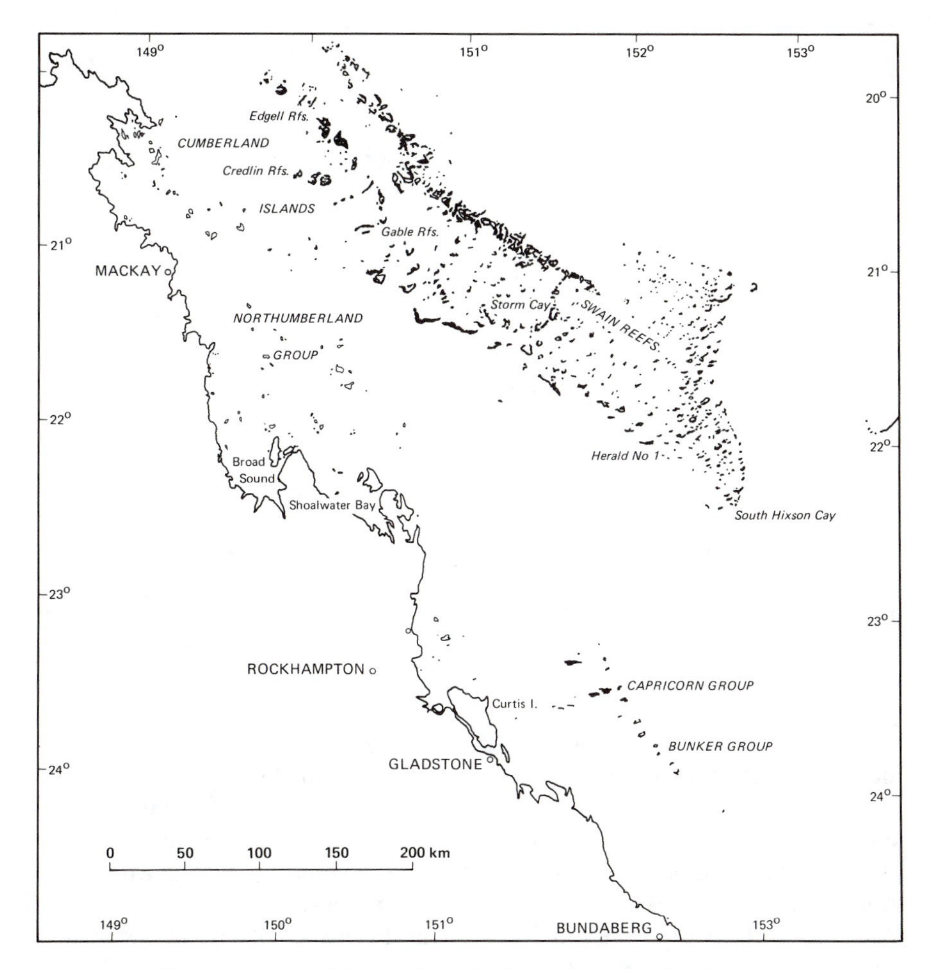

Figure 8.2 Map of the Southern Region of the Great Barrier Reef

tal origin. To the north of the Swains group the outer reefs of the Southern Region form a tightly packed mass of large reef separated by narrow channels that break the force of the Pacific Ocean waves.

The Central Region (Fig 8.3) extends northward from 21°30'S to approximately 14°30' S. The continental shelf and the inshore lagoon become gradually narrower. By 15° the lagoon is less than 10 nautical miles wide and the continental shelf less than 30 nautical miles. The depth in the lagoon and between reef masses gradually decreases to the north to be about 35 meters at 15°S. Three major river systems drain into the lagoon between 17° and 21°. Up to about 17°S the reefs are relatively sparsely scattered and there is no distinct outer line of reefs abutting the Pacific. From 17° northward, however, the ribbon reef system stands on

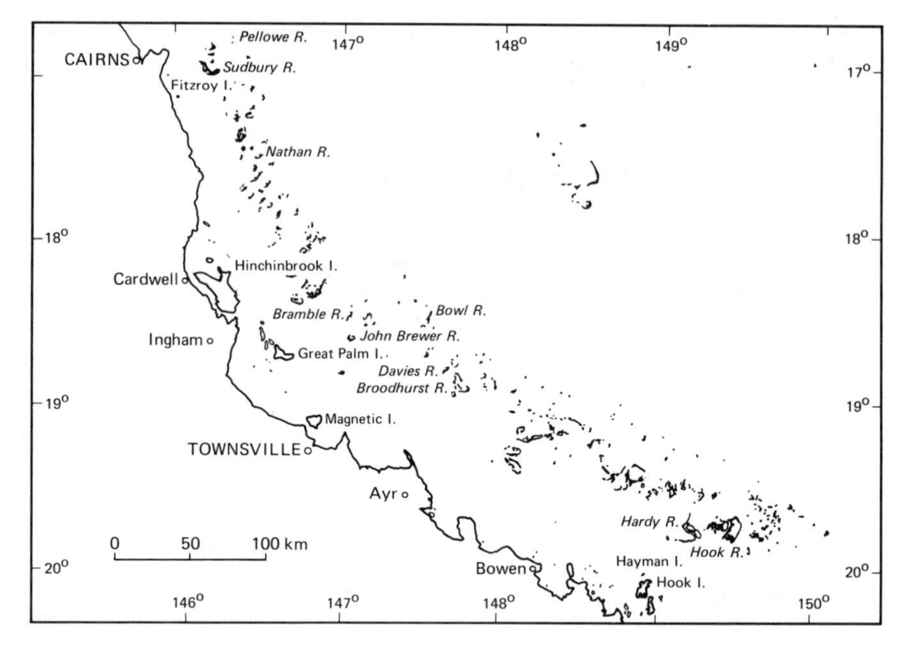

Figure 8.3 Map of Central Region of the Great Barrier Reef

the edge of the steep continental shelf dropoff forming an almost continuous rampart of coral broken only by narrow channels.

The Northern Region (Fig 8.4) runs from about 14°30' S to the Torres Strait at approximately 10°S. It has little or no inshore lagoon and the depth is for the most part about 30 meters. The continental shelf becomes broader again and no major river systems drain into the region until in the extreme north, where the influence of the Fly River of Papua becomes a factor.

The Great Barrier Reef consists of some 2,900 individual reefs and 250 cays. The reefs range in size from less than a hectare to the massive structure of Cockburn Reef in the Northern Region, which covers 30,130 hectares. There is an immense variety in the structure and geomorphology of the individual reefs. Hopley[3] identified six stages in the development of shelf reefs and three of ribbon reefs.

The Great Barrier Reef is a living system of biogenic reefs composed of the accumulated limestone skeletons of corals, algae and other reef animals and plants. The spaces between the large skeletal pieces are filled in with smaller fragments in the form of sand and gravel. The surface layer is often cemented by a thin hard layer of living coralline algae, whereas within the reef mass the accumulated material is gradually cemented by chemical changes over time. Much of the surface of the reefs supports en-

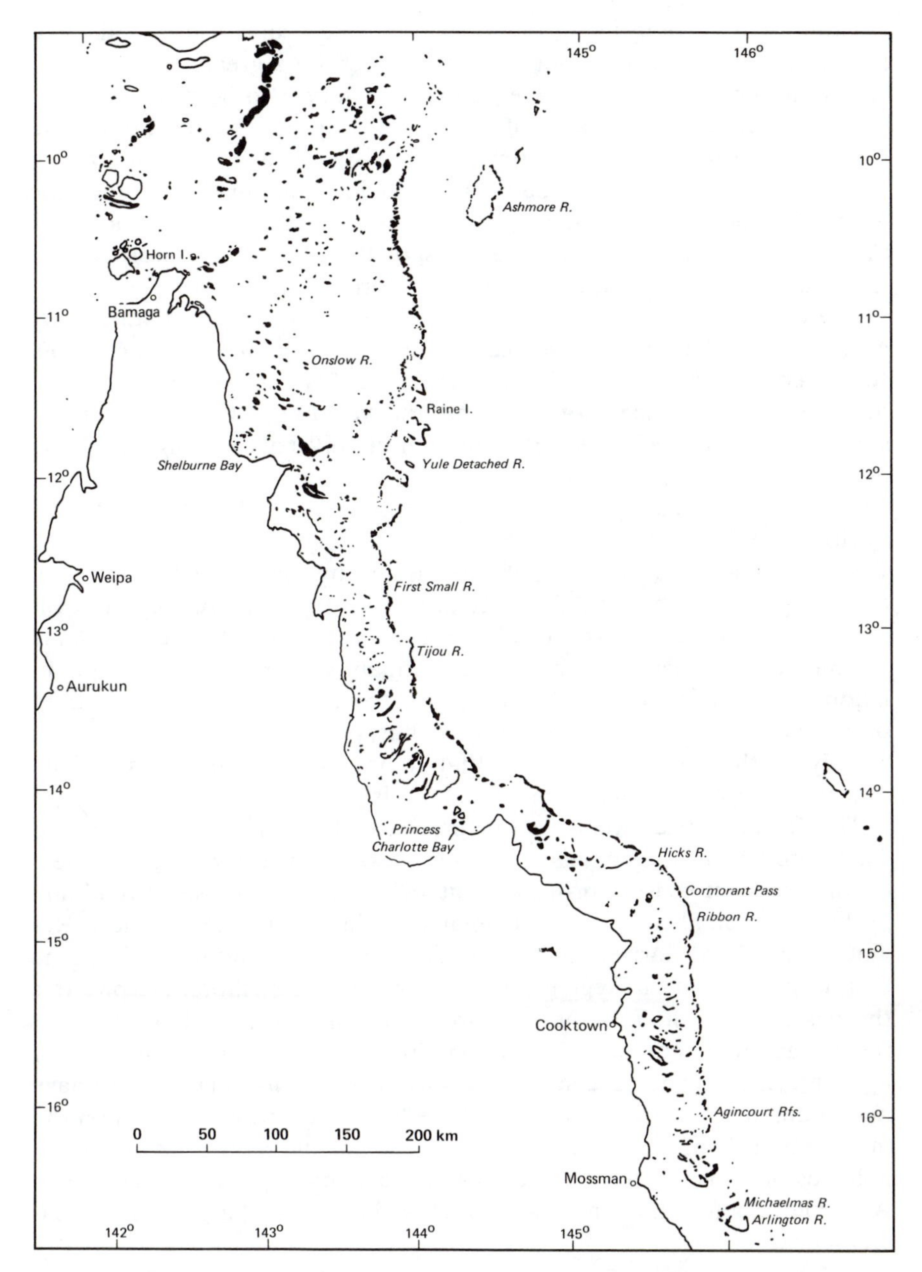

Figure 8.4 Map of the Northern Region of the Great Barrier Reef

crusting or branching living biota such as algae, corals, soft corals, and sponges, which in turn provide shelter and food for a great diversity of invertebrate animals and fish. Even areas that appear to be dead or uncolonized are covered by a layer of microscopic algae or bacteria that form the basis of a food chain for grazing organisms. It is a dynamic system, built up and maintained by the addition of limestone mainly through the growth of coral and coralline algae. It is weakened and broken down by the erosion of boring algae and sponges, by the burrowing of worms, molluscs, and crustaceans, and the action of storms.

The Great Barrier Reef has 71 genera of hard coral. The great majority of the species that occur there occur elsewhere in the Indo-Pacific region. In biogeographic terms the Great Barrier Reef is a major part of the Indonesian-West Pacific region which has the world's greatest genetic diversity of reef species.[4] The species and generic diversity are highest in the northern Great Barrier Reef and decrease gradually toward the southern end. This is, for example reflected in the taxonomic study of the family Poritidae by Veron and Pichon.[5]

The scale and diversity of habitats offered by the Great Barrier Reef and the great number and diversity of species that they support make it one of the world's most spectacular marine ecosystems. Its location some distance off the coast of an area of a continent with a small human population has meant that it has been exposed to very low levels of pollution[6] or use. Despite historic fisheries for trochus, pearls, and beche de mer[7], it is only in the past few decades that it has been subject to sustained fisheries[8] or even substantial numbers of visitors.[9]

The Great Barrier Reef lies off the coast of the state of Queensland. There are 12 towns or cities and a number of smaller population centers on the adjacent coast. North of about 16°S, there is no coastal road and nothing but small scattered settlements on the coast. The Northern Region of the Great Barrier Reef is thus remote and difficult to reach. In the Southern and Central Regions, most reefs are well offshore. Despite this the resources of the nearshore environment and reefs are important for the recreational activities of their communities and for the population of the hinterland. They have supported local fisheries as long as there have been human settlements, although these have been based more upon the inshore and mangrove species than on those of the reefs. The new technologies of transport have made offshore reefs more generally accessible. As a consequence, the ranges of recreational and tourist activities have increased.

HISTORY OF RESEARCH

Information on Australian Aboriginal use of the sea is poor. The Great Barrier Reef Marine Park Authority held a workshop to review traditional knowledge of the marine environment of northern Australia. That workshop[10] indicated that there is good evidence of long use of the coastal and marine environments by Aboriginal people. There are substantial middens in Princess Charlotte Bay that have been dated indicating extensive use of shellfish and dugong 5,000 years before the present. The scarcity of information is not particularly surprising since sea level only stabilized at its present level between 8,000 and 6,000 years ago. So, although there is good archaeological evidence of Aboriginal communities inland up to 40,000 years before the present, much of the earlier coastal remains would have been covered by the rising sea. Early European explorers and anthropologists noted the use of a variety of styles of canoes and a wide range of fishing techniques including nets, hooks, spears, and traps.

The Great Barrier Reef has had a special role in European and colonial history. The Torres Strait at the northern end was known to Dutch and Portuguese navigators from the end of the sixteenth century as a sea route from the Dutch settlement of Batavia, at Djarkarta on the island of Borneo, to China and the Far East[11]. The Great Barrier Reef to the south was little known although maps of Chinese origin existed. The near disaster encountered by Lieutenant James Cook in 1770 brought the Great Barrier Reef spectacularly to the attention of the scientific world. For Cook, the Great Barrier Reef was a fearsome obstacle, "a monstrous labyrinth" of sharp reefs on a lee shore. His vessel *HMS Endeavour* ran onto a reef of the Great Barrier Reef. The heroic and romantic account of how the ship was saved by skilled seamanship, taken into the estuary of what is now the Endeavour River, careened and repaired before continuing to complete the expedition is contained in the journal of his 1768–1771 expedition. Sponsored in part by the Royal Society of London, the records of that expedition contain the earliest scientific reports of the Great Barrier Reef.[12]

Following Cook's exploration, in 1788 Great Britain established a colony, New South Wales, in the southeast of the Australian continent. Within 50 years much of the south east had been explored; valuable grazing, agricultural land and timber had been discovered and many settlements as far north as Brisbane had been established. The colony grew at a time when Great Britain was at war with France. It became increasingly apparent that supply, security and survival of the colony would depend upon better communications wlth the major British bases in India. Future growth and the marketing of the products of the colony would also depend increasingly upon developing alternatives to the long and hazardous shipping route across the Great Australian Bight and through Bass Strait to the south of the continent.[13]

Scientific research and exploration of the Great Barrier Reef was thus first driven by the need to identify a safer shipping route to the East Indies and thus to Britain's major Far East colony, India. Voyages of survey and exploration by Flinders, Jukes, Darwin, and others saw shipping routes identified and surveyed and the geology and geography of the Great Barrier Reef described.[14] As the new shipping routes were used, European settlement of the continent moved northward to make use of the marketable natural resources and to develop the land for agriculture. On the Great Barrier Reef, the phosphate-rich guano deposits of the islands of the Capricorn Bunker group in the south and of Raine Island in the north were mined. Sea turtles were harvested on and around these and other islands as they came there to mate and to nest. The quest for further economic opportunities led to the systematic study of possible fisheries and marine resources by Saville Kent.[15]

Early research was conducted by expeditions of some weeks or months duration generally ranging widely in the Great Barrier Reef. These revealed opportunities for fundamental studies of natural processes in an environment little touched by human activity. Biologists saw the potential for studying the interactions of animals and plants in complex and diverse communities. Geologists saw the potential to study the structure and age of the reef and the processes by which corals reefs accumulate limestone, and limestone accumulates into reefs. The first detailed study and data collection program at a specific site over the course of a year did not occur until a prolonged study expedition in 1928–29 supported by the Royal Society of London and the Great Barrier Reef Committee.[16]

Permanent research facilities developed slowly on the reef and adjacent mainland. In 1958 the Great Barrier Reef Committee established a research field station on Heron Island with an aquarium and laboratory space for visiting researchers.[17] This provided a permanent expeditionary base that enabled visiting researchers in many disciplines to conduct projects on the reef. Some, for example, Connell[18] established long–term studies based on repeated visits to permanent sites. Otherwise, long-term studies were the province of graduate students able to spend protracted periods at the station. The first facility designed specifically for permanent onsite study was a small fisheries research laboratory established on Green Island in 1965 to provide for continuous on-site investigations into the extent, impact and behavior of the coral-eating crown of thorns starfish.[19] That laboratory still exists, but its permanent use ceased in 1969 when operations transferred to Mourilyan Harbour near Innisfail following the collapse of the Green Island crown of thorns starfish population.

In 1962 the first step toward development of major permanent facilities for reef research and training in North Queensland was taken following the decision that the university being set up in Townsville should establish a Department of Marine Biology. That department was established in 1967 and by 1972 James Cook University of North Queensland had well-equipped research laboratories in Townsville and a research vessel capable of working anywhere in the Great Barrier Reef region. In 1974 research capability was further boosted by the establishment near Towns-ville of a major government-funded facility, the Australian Institute of Marine Science. By the end of the 1970s there were four permanent research stations, at Lizard Island in the Northern Region, Orpheus Island in the Central Region, and Heron and One Tree Islands in the Southern Region.

Chapter 9

The Great Barrier Reef Marine Park: The Campaign For Management Of The Great Barrier Reef

---

The Great Barrier Reef Marine Park is apparently the first case of coordinated multiple use management of a marine environment on a scale consistent with marine processes. It is exceptional in several ways.

The establishment of management preceded and anticipated many potential problems. The Great Barrier Reef is a natural environment of a technologically advanced nation, yet until the 1960s it was generally inaccessible and little touched by human impacts. Despite this it was known to be an area of global environmental importance and, as such, an object of national pride. The passage of legislation and the establishment of a management system occurred as problems were starting to arise and not as a belated response to acute problems.

In the course of the 1950s and 1960s, strong markets for sugar and beef, the major primary products of North Queensland, led to growth of the coastal communities adjacent to the reef. The 1960s were a time of economic change in the state of Queensland. A conservative government first elected in 1957 was strongly entrenched with a mandate to reduce the gap in living standards between Queensland and the states to its south. Economic development was the priority concern of the government.[1] A program of oil exploration in the Great Barrier Reef was established. Mining of reefs for limestone to enable the manufacture of agricultural lime was proposed. A massive increase in fisheries was foreshadowed as was the development of a major tourist industry based on recreational and sport fishing on the reef.

With the development of a broader economic base for Queensland, able to support and maintain improved technologies in boat building and in motors for boats, the Great Barrier Reef began to be accessible. Film, television, and magazines brought its wonders directly to the attention of people throughout Australia and the world. Initially it was seen as a rich, diverse, and virtually untapped source of fish for recreation, sport and commerce. The development of reliable diving gear and underwater cameras brought an entirely new perspective on the reef. The variety of form and color of the fish and the intricacy and delicacy of the corals made the coral reefs generally, and the Great Barrier Reef in particular, striking examples of beautiful and fascinating natural environments vulnerable to misuse and abuse by humans.

The 1960s was the decade in which environment protection and conservation became issues of great public and political interest in the wake of rapid economic development in Western Europe and North America. Concern over human impact on the environment, and consequent impacts on human health and the quality of human life became a major issue internationally following the work of authors such as Carson and Commoner.[2] As the United States and European economies grew, and as the results of research into environmental issues were publicized, the negative consequences of unrestrained economic development became increasingly apparent. This resulted in rapid membership growth of environmental groups in most developed countries. Where, historically, these groups had focused primarily upon protection of natural areas for research and recreational appreciation of nature, they expanded their activities to reflect deeply held concerns about the survival of natural environments. Typically, membership covered a wide spectrum of political opinion.

The same issues were taken up in Australia, and the Great Barrier Reef became both a topic and a symbol, reflecting local issues, deeply held political philosophies and contemporary international awareness of the fragility of the environment. In Australia the Australian Conservation Foundation was established in 1965, the Wildlife Preservation Society of Queensland in 1963, and the Australian Littoral Society (Queensland) in 1967. These groups developed a public campaign to save the Great Barrier Reef. The various factors that contributed to the campaign are discussed briefly here. They have been recorded and discussed by Clare[3] and, with the insight of one who was actively involved in most of the political struggles over the Great Barrier Reef, by Wright.[4]

As the largest coral reef system in the world, the Great Barrier Reef was regarded widely as one of the wonders of the natural world. Australians generally felt national pride and a responsibility of stewardship, yet the Reef was little researched and virtually unprotected by law. Local, national and international concern developed over the need for conservation legislation to protect the Reef from human threat. The issues of potential threats to the Great Barrier Reef were viewed with increasing concern by the environmental movement since the Queensland Government was generally very hostile to anything which might discourage, delay or limit economic growth, such as proposals for review, amendment or rejection of development proposals in order to limit environmental impact.

Key players in the developing controversies were the Wildlife Preservation Society and the Australian Littoral Society[5] which became increasingly concerned as the Great Barrier Reef was threatened by mining with little or no apparent concern for the environmental impacts. Another key player in the developing controversy was the Great Barrier Reef Committee, a respected and expert group of scientists and others established in 1922 to promote scientific research on and to protect the reef.[6] The committee was initially less concerned than the conservation groups but as potential hazards to the reef increased and were apparently ignored or lightly dismissed by the Queensland government, it became seriously concerned. As a group and individually its members began to work with conservation groups.

The Wildlife Preservation Society saw the threat of a dangerous and far-reaching precedent in an application in 1967 for a mining lease to take coral limestone from Ellison Reef off the coast near Innisfail. The application concerned relatively small amounts of limestone to be used for the production of agricultural lime. However, there was talk of expanding the Australian steel industry with consequent greatly expanded demand for limestone. Strong reaction to the proposal came from conservation groups and scientists who appeared at the mining warden's court hearing to present evidence opposing the grant of such a lease. After a widely publicized hearing, the application was refused and the warden's decision was accepted by the Queensland Minister for Mines. Despite this, those opposed to reef mining were concerned by the powers of the Queensland Minister for Mines to overrule such a decision. To those who were skeptical of the commitment of the Queensland Government to effective conservation and protection of the Great Barrier Reef, the case highlighted the need for legislation at a higher constitutional level.

The central technical controversy over the oil drilling issue was whether exploration and production could be conducted without having significant adverse impacts on the environment. There were two components: the reliability of the various technologies for preventing spills, and the extent of any hazardous impacts that might arise in the event of a spill.

The seriousness of the consequences of a spill was far from clear and was the subject of heated and emotional campaigns with all parties able to call upon some scientific evidence to support their views. However, there was so great a range of assumptions and techniques behind the experimental designs used to collect the evidence that objective evaluation required a major effort.

The controversy grew. On one hand it was claimed that an oil spill at the south of the Great Barrier Reef would have impacts that would be carried northwesterly along the reef by the prevailing southeast trade winds and would thus cause enormous damage to most of the reef and have consequent severe economic impacts on coastal communities, fishing, recreation, and the developing tourist industry. On the other hand, it was claimed that the risks could be so greatly reduced by proper management practices that the level of public concern was unreasonable even hysterical.[7]

As the controversy developed, seismic studies were carried out by the industry and six wells were sunk in the 1960s, including three in 1968 in the Swains group of reefs at the southern end of the Great Barrier Reef. Accidents elsewhere on the Australian continental shelf, in Bass Strait and in Buonoparte's Gulf, and a series of major accidents with marine oil platforms overseas, e.g. Santa Barbara, California, and the North Sea, indicated the environmental problems associated with the routine of hydrocarbon exploration and production. Accidents with oil tankers, e.g. *Torrey Canyon* on rocks off the south west coast of England in 1967 and *Oceanic Grandeur* on an unsurveyed deep shoal in the Torres Strait indicated the hazards of additional tanker traffic in the dangerous waters of the Great Barrier Reef shipping lanes.[8]

In 1970 a moratorium was declared on further drilling and the Commonwealth and Queensland governments established and jointly funded a three man expert group to serve as conjoint Royal Commissions to inquire into the issue of exploratory and production drilling for petroleum hydrocarbons in the Great Barrier Reef region. The summary and recommendations of the Royal Commissions were not released until late 1974 and their report until 1975,[9] but they did not resolve or reduce the controversy. The majority recommendation was that there were some areas of the Great Barrier Reef region in which petroleum exploration could be conducted under stringent conditions without unacceptable risk. The chairman in a minority opinion dissented from his colleagues, maintaining that no exploration or production should take place since the risks inherent in the technologies examined by the commissions were too great.

The issues of oil drilling and limestone mining were critical tests of the resolution and ability of governments to manage and conserve a spectacular environment in the face of perceived threats arising from pressure for economic development. In both cases there was scientific debate but no satisfactory conclusion regarding the extent and significance of perceived threats arising from the activities in question. It was clear that scientific understanding of the reef ecosystem was very limited and that research was urgently needed to address many of the implied questions. In the meantime conservationists and scientists saw the need for caution and conservatism in assessing potential impacts before establishing new activities or expanding existing ones.

Concern at possible human impacts upon, and lack of understanding of, the reef ecosystem were further fueled by the occurrence of large populations of the coral eating crown-of-thorns starfish *Acanthaster planci.* First reported in 1965, the effect of such populations was dramatic and immediately devastating with their predation killing as much as 95% of the living coral on severely affected reefs. This, too, became a controversial issue since its cause was unknown and there were several reasonable avenues of speculation that divided scientific opinion on the degree of probability that it was man-induced, man-enhanced, or a natural phenomenon independent of human impacts. There were widespread calls for government action, including research to enable better understanding of the phenomenon, banning the collection of the triton shell (a predator of the starfish), and mobilizing volunteers, the defense forces, or specialist teams to collect and destroy adult starfish. There was also widespread controversy. Those who inclined to the view that the phenomenon was natural suggested that general controls would be unwarranted and possibly damaging interference in ecological and evolutionary processes. Those who held that the phenomenon was caused by human activity saw controls as essential to saving the Great Barrier Reef.

The scientific controversy was reviewed by Potts,[10] and following the start of a second occurrence by Moran.[11] The public and political responses to both episodes of the phenomenon were reviewed by Kenchington.[12] The central debate as to whether it was natural, man-induced or man-enhanced led to debate regarding the seriousness of the impact in relation to the dynamics of reef ecology and other forms of coral damage. This in turn led to debate over the need for controls. Despite two decades of research there are still no generally agreed hypotheses regarding its causation and no clear guidelines for management of the phenomenom.

Despite enthusiasm for controls, they were costly and not very effective. No method of control had been demonstrated other than treatment of individual starfish. This often involved some damage to corals as divers searched for and removed or treated individuals hidden at the base of branching coral colonies. The populations of starfish that were causing problems contained tens or hundreds of thousands of individual animals. A large proportion of these would be hidden in crevices under corals or boulders. Even in the densest populations where there was no need to search for starfish, divers could treat relatively small numbers. Figures for sustainable average rates quoted for an experimental control program showed that a diver could collect only 38 starfish per hour and even with the most efficient treatment, injecting toxic chemicals, could treat no more than 132 individuals per hour.[13] Management therefore focused on tactical controls, reducing starfish numbers at sites of specific interest in order to protect corals for the purposes of tourism, recreation, or research.

In the context of the debate of the early 1970s regarding management and protection of the Great Barrier Reef, the crown-of-thorns starfish controversy demonstrated the scale and difficulty of management problems, the need for scientific research, and the need for an agency to co-ordinate reef management. A "Save the Reef" campaign started on the basis of concerns about the crown-of-thorns starfish assumed a much wider signi-ficance as a general campaign for reef protection and management.

In addition to the issues raised by potential mineral extraction and the uncertainties of the crown-of-thorns starfish situation, some more traditional marine resource management issues were surfacing. A profitable trawl fishing industry became established in the 1950s using small vessels operating out of North Queensland ports working resources of prawns and scallops on the continental shelf. It was clear that larger vessels equipped with freezers could operate farther afield and that there were large potential export markets in the United States, Japan, and Europe. By the late 1960s the trawl fishery was expanding rapidly; there were still large areas that had not been surveyed; and there was potential for further growth. Nevertheless, some of the longer established operators saw the deterioration of the earliest discovered grounds and suggested that it would not be long before management would be needed to avoid overexploitation of the stocks.

The need for some kind of action was more urgently recognized by fishermen when foreign fishing vessels began to use Great Barrier Reef waters. Most of these vessels were Taiwanese. Two types of operation were involved. One was trawling using modern vessels much larger than those of local fishermen, who claimed that, on occasion, the Taiwanese shadowed local vessels and then trawled alongside them when they had located productive areas. Some fishermen and conservationists were shocked to find that large Taiwanese vessels would sometimes work in pairs using heavy chains to clear areas of the seabed of corals, sponges and other benthic cover.

The other Taiwanese fishery involved old-style vessels that visited the Great Barrier Reef to collect giant clams, turtles, reef fish, and ornamental shells. Fishermen and charter boat operators were particularly incensed by the giant clam fishery, which involved obvious destruction to reef top areas. This matter was addressed in part by protection of sedentary fauna through the Continental Shelf (Living Natural Resources) Act 1968. Nevertheless, the need for more comprehensive management of the Great Barrier Reef and reef region fisheries was clear.

Another factor was the potential for development of a major tourist industry on the Great Barrier Reef. For some years the reef had been one of the international marketing attractions of Australian tourism. This had increased in the 1960s, probably as a consequence of films and television documentaries, although with the exception of a few sites the reef was virtually inaccessible to all but the wealthiest and most leisured visitors.[14] Yet the potential was clear and the experience of the Caribbean and Mediterranean tourist development booms was not reassuring. Tourism, properly planned and managed, by creating an economic value for the natural reef environment, could be an ally of conservation. Uncontrolled tourist development could lead to damage to the reef, damage and destruction of related coastal environments, and resource competition between old and new users.

The issues had wider national implications because of differences between the Australian Commonwealth and Queensland Governments concerning the constitutional interpretations of jurisdictional responsibility for the Great Barrier Reef. The concern of a majority of Queensland citizens and of those of the rest of Australia was that the Queensland Government would not protect the reef in any decision that involved foregoing or significantly restricting development.[15] For the Commonwealth the issue of the environment generally, and of the Great Barrier Reef in particular, became increasingly prominent politically. Although there were differences in approach, the electoral platforms of both political blocs at the Commonwealth level contained policies for environment protection. This was particularly significant because implementation of such policies was almost certain to bring federal (Commonwealth) legislators into situations in which the national significance of environmental matters would create strong conflict with traditional concepts of state jurisdiction.

For the Queensland Government and for other states of Australia the constitutional issues were of critical significance. It was claimed that the potential wealth that would flow from discovery and development of significant oil fields under the Great Barrier Reef could be a major factor in transforming the state's economy from an agricultural to a broader base. If the powers of the Commonwealth could be used unilaterally to block this potential, the independence of the states would be eroded.

The outcome of the various controversies was a widespread view that new Commonwealth legislation was needed to provide for the management of the Great Barrier Reef. The desirable form and objectives of such legislation were less widely agreed. Options ranged from minor amendments to fisheries legislation to special legislation to make the Great Barrier Reef a national park. In the early 1970s, it was apparent that the next Commonwealth election would be closely fought. Specific legislative proposals for the Great Barrier Reef were developed by a number of nongovernment organizations which sought successfully to have them included, in full or in part, in the policy documents of the various political parties that would contest that election.

Chapter 10

The Great Barrier Reef Marine Park—The Constitutional Setting

The Great Barrier Reef lies on the continental shelf of Northeast Australia off the coast of the State of Queensland which is one of the six states that comprise the Commonwealth of Australia, a federation of states formed in 1901. The Commonwealth constitution defines certain powers, including foreign affairs and defense as areas of federal jurisdiction in which Commonwealth legislation takes precedence over state legislation covering the same field.[1] The constitution contains a provision for a state to refer specified powers to the Commonwealth and also provides referendum procedures for constitutional amendment, which may involve the transfer to the Commonwealth of jurisdiction over additional powers. There is thus a range of matters in which the Commonwealth may, under the Constitution and subsequent legislation, make laws that take precedence over state laws. However, if Commonwealth Law is silent on such a matter, state Law may constitutionally fill the vacuum.

Under the constitution, the states retained responsibility for regulation of internal matters relating to the territorial sea adjacent to their mainland or island territory. Only in the case of waters adjacent to land added to Australia since the formation of the Commonwealth and of certain lands transferred to or acquired by the Commonwealth did the Commonwealth have internal responsibility for management of the Australian territorial sea.

Beyond the three-mile territorial sea, the Commonwealth of Australia established and exercises rights and accepts responsibilities concerning the living and nonliving resources on the continental shelf adjoining its shores to a depth of 200 meters or greater to the limit of exploitability. This occurs under the terms of the 1958 Geneva Convention on the Continental Shelf to which Australia is a signatory. This led to the enactment of the Petroleum (Submerged Lands) Act 1967 and the Continental Shelf (Living Natural Resources) Act 1968.

Australia adheres to a territorial sea extending three miles from low water. It is clear that a large part of the Great Barrier Reef lies on the continental shelf but beyond the three-mile territorial sea. It is also clear that most of the actively used intertidal areas come under the jurisdiction of Queensland. At the time the controversy over the Great Barrier Reef was at its height, there was a range of legal opinion concerning the extent to which the Commonwealth could control subtidal areas within the three-mile territorial sea of the mainland or islands under the jurisdiction of Queensland.

In 1973 the Commonwealth Parliament used its constitutional powers in the area of international relations to provide for fuller implementation of the 1958 Geneva Conventions by enacting the Seas and Submerged Lands Act 1973. This established that the boundary of the jurisdiction of the Commonwealth was the low water mark. The legality of this act was challenged in 1974 by the state governments of Queensland and west Australia, but its validity was upheld by the High Court of Australia.

The effect of the Seas and Submerged Lands Act 1973 was thus to place the subtidal areas of the Great Barrier Reef under the jurisdiction of the Commonwealth. Nevertheless intertidal areas attached to the mainland, or to an island forming part of the state, remained under the jurisdiction of the state.

In a situation where an island could form the focus of a claim of state jurisdiction over part of the Great Barrier Reef there were immense possibilities for jurisdictional challenge. A major problem was to determine which islands were part of Queensland at the time of the formation of the Commonwealth in 1901 and which had been added since federation. One source for this information was the letters patent issued when the State of Queensland was created from the northern part of the state of New South Wales in 1859. The letter's patent referred to islands but contained no listing of them. Few of the islands appeared on contemporary maps and even fewer had been surveyed.

The situation of inaccurate or nonexistent historical mapping is not unusual, but the nature of coral reef islands gives rise to unusual complexities in determining the precise location of boundaries between the jurisdictional responsibilities of the State of Queensland and the Commonwealth of Australia. Reef islands exist in a highly dynamic environment. They may form over periods of decades by the gradual accumulation of reef debris, but they can disappear or change their location, either suddenly in a severe storm, or gradually following a change of strength or direction of prevailing wind or current over a long period. A well-documented case is a cay on Wheeler Reef in the central region of the Great Barrier Reef, which protrudes above high water during the season of the southeast trade winds but is flattened and may be moved southward during a strong season of northerly winds to such an extent that it is only exposed at midtide levels.[2] The matter of whether similar islands in more remote parts of the Great Barrier Reef region existed at the time of creation of the State of Queensland would be very difficult to determine.

A second problem is to determine the location of the low water mark at many sites where the profile of sandy beaches or the position of sand islands may change by hundreds of meters in a short period of time with the change of prevailing winds between seasons or as a result of a severe storm. This situation is further complicated by differences of opinion between Commonwealth and Queensland Governments regarding both the definition of low water and the position of low water in relation to an island on a coral reef.

The Commonwealth defines low water as the mean low water; that is the mean value over a period of time, usually a one month lunar tidal cycle. The State defines low water as the low of Indian Springs. Figure 2.1 in Chapter 2 illustrates the relationship of these definitions.

Regardless of the definition of low water, there is further difference of opinion. The Commonwealth holds that the position of low water around an island is the position at which standing water is encountered at the time of low water. Queensland holds that it is the outermost part of the structure attached to the island that is exposed at low water. Since most reef flats occur at or about mean low water, the difference could be critical (Fig. 10.1).

The situation was undoubtedly complex. A test case could be taken to

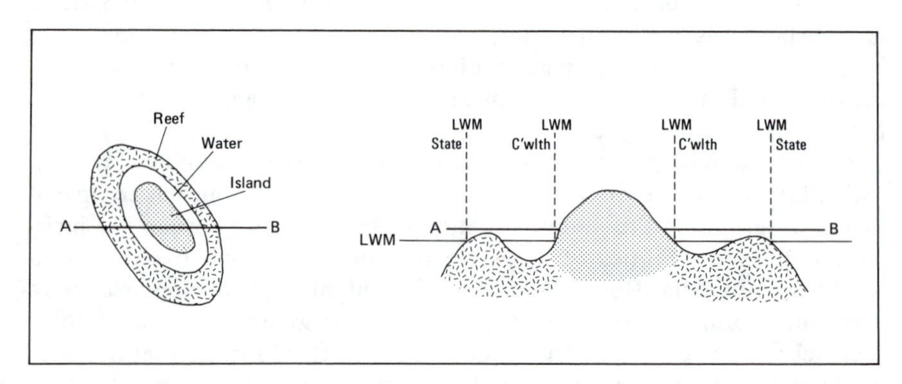

Figure 10.1 Diagram showing plan and section of an island surrounded by reef, illustrating different interpretations of Low Water mark.

the High Court to decide between alternative definitions of low water, but it was apparent that even with an agreed definition there was potential for dispute over its application at almost any location with a sandy beach in a physically dynamic environment. Further, it was increasingly recognized that whereas there had to be a boundary for constitutional and jurisdictional purposes whatever boundary was established there would be important elements of the regulation of human activities and the reef ecosystem that crossed it. A graphic example is the case of sea turtles, which hatch from nests on land under Queensland jurisdiction, move to the sea across intertidal areas under state jurisdiction, cross the low water mark to enter Commonwealth jurisdiction, and then move on to feed and grow for years in international waters. Eventually they return to the Great Barrier Reef to mate in areas under Commonwealth jurisdiction and for the females to lay eggs on Queensland territory.

In simple terms the outcome of the UNCLOS 3 was that it is generally accepted that a nation has the right to make provision for the regulation of activities relating to use and conservation of the resources of its continental shelf and in a zone extending from its territorial sea to a line 200 miles from its baseline. This right does not entitle nations to restrict the innocent passage of foreign vessels, or scientific research or to deny reasonable access to resources that it does not exploit itself.

The essential issue for those concerned for the protection of the Great Barrier Reef was that there needed to be an arrangement that could provide for the management of all the ecosystems of the Great Barrier Reef region and that could resist local political pressures. UNCLOS 3 provided a basis for more comprehensive Australian legislation to conserve and manage the Great Barrier Reef.[3]

There was a mood among some of those campaigning for the establishment of a national park on the Great Barrier Reef that the Commonwealth should override or ignore the protests of the Queensland Government. Nevertheless it was clear there was no legal basis for this to be done. The Commonwealth did not have jurisdiction over representative supratidal ecosystems. Neither, despite the passage of the Seas and Submerged Lands Act 1973, did it have jurisdiction over most of the distinctive intertidal areas adjacent to the coast of the mainland and islands that formed part of Queensland.

Jurisdictional responsibility for the Great Barrier Reef is clearly divided between the Commonwealth and the state of Queensland. Any logical approach to management of the Great Barrier Reef has to involve both Governments. In the event the Commonwealth's Great Barrier Reef Marine Park Act 1975 was designed to provide for extensive ivolvement of the state in the decision-making process.

THE GREAT BARRIER REEF MARINE PARK ACT

The unanimous passage by the Commonwealth Parliament of the Great Barrier Reef Marine Park Act in June 1975 marked the culmination of a long public campaign and a difficult legislative drafting process needed to address the complex jurisdictional and political issues discussed in chapter 2.

The object of the act is defined in Section 5(1): "to make provision for and in relation to the establishment, control, care and development of a marine park in the Great Barrier Reef Region."

As with most such legislation, a major part is concerned with arrangements for the establishment, control, and management of the executive agency to undertake the work necessary to achieve the object. The features that provide major keys to the approach to planning and management are described briefly here.

The boundary of the Great Barrier Reef Region is defined in a schedule to the Act as commencing at the northernmost extremity of Cape York, running due east to a point well to the east of the continental shelf and thence through a series of points to the east of the Great Barrier Reef until it reaches latitude 22° 30'S which is south of the Great Barrier Reef, where it runs due west until it intersects the coastline of Queensland at low water. It then runs northerly along the coastline at low water back to the point of commencement (Fig 10.2.).

The boundary was drawn in accordance with the Seas and Submerged Land Act 1973 and the definition refers to that act in excluding from the Region any island or part of an island, that forms part of Queensland and is not owned by the Commonwealth.

The Act established the Great Barrier Reef Marine Park Authority and in S 7 (1) defined its functions:

(a) To make recommendations to the Minister in relation to the care and development of the Marine Park including recommendations, from time to time, as to
(i)The areas that should be declared to be parts of the Marine Park
(ii)The regulations that should be made under this Act.

(b) To carry out, by itself or in co-operation with other institutions and persons, and to arrange for any other institutions or persons to carry out, research and investigations relevant to the Marine Park.

(c) To prepare zoning plans for the Marine Park in accordance with Part V

(d) Such functions relating to the Marine Park as are provided for by the regulations.

(e) To do anything incidental or conducive to the performance of any of the foregoing functions.

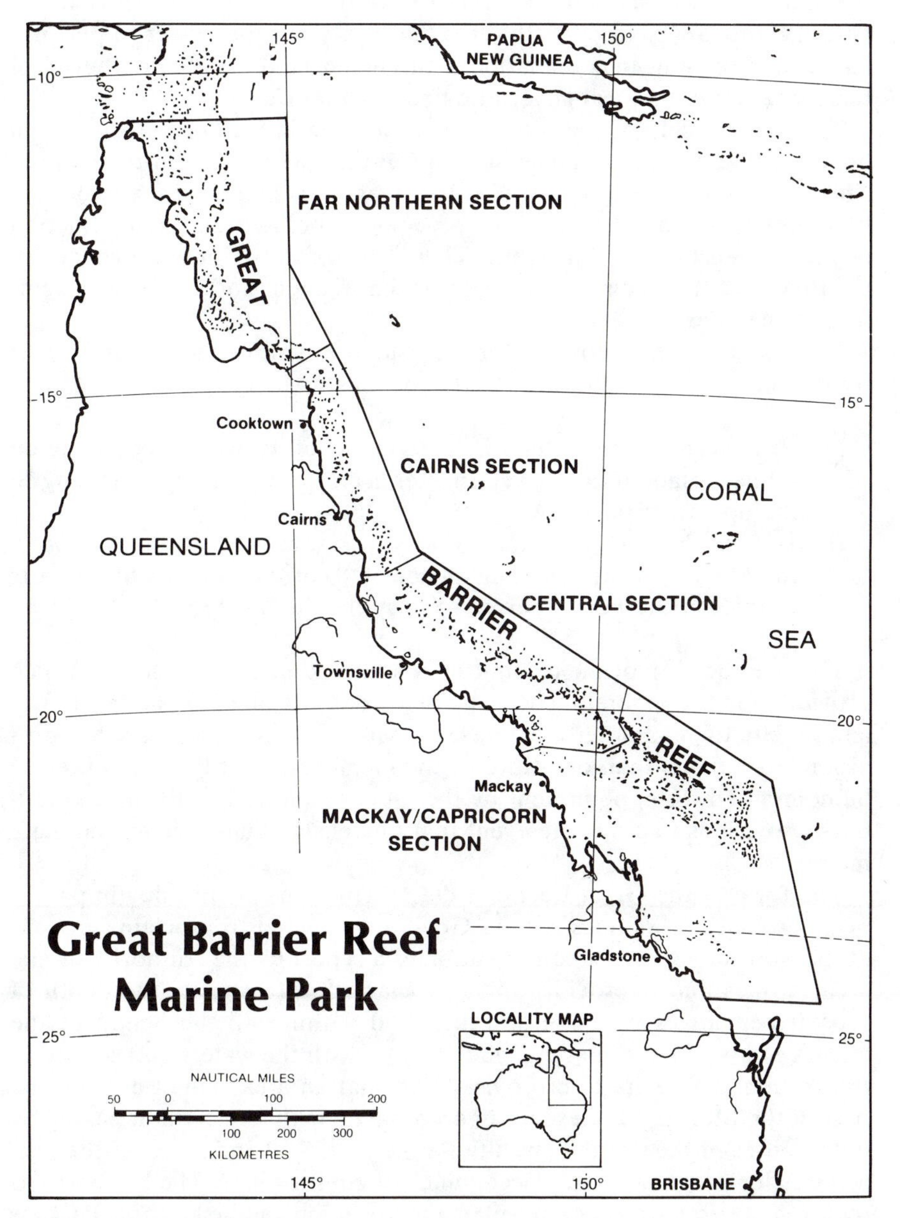

Figure 10.2 Map showing the Great Barrier Region and the four sections of the Great Barrier Reef Marine Park.

The wording of the Act makes several provisions for the involvement of the State of Queensland in the processes of the Marine Park. Thus, in the first of these, S8(3) specifies that the Authority has the power to perform any of its functions in cooperation with Queensland, with an authority of that state or with a local governing body of that state.

The membership of the Authority is specified in S 10 of the Act which provides that the chairman shall be a full-time member with the responsibilities of chief executive. It also provides that there shall be two part-time members, one of them to be a person appointed on the nomination of the Queensland Government. There are also provisions to cover appointment of that member in the event that Queensland declines to nominate a member.

Part IV of the act provides for the establishment of the Great Barrier Reef Consultative Committee. S21(1) defines its functions:

(a) To furnish advice to the Minister, either of its own motion or upon request made to it by the Minister, in respect of matters relating to the operation of this Act.
(b) To furnish advice to the Authority in respect of matters relating to the Marine Park, including advice as to areas that should be parts of the Marine Park referred to it by the Authority

The membership of the committee is specified in S 22 of the Act which provides for the committee to consist of one member of the Authority and no less than 12 other members. As with the Authority, there are provisions for no less than one-third of the membership of the committee to be nominated for appointment by the Queensland Government and fall-back provisions to cover the event that Queensland declines to nominate members.

The form of the Great Barrier Reef Marine Park is specified in part V of the Act. S 31 provides for the Governor-General to proclaim sections of the Marine Park after consideration of a report by the Authority on the area in question. It provides for a proclamation to specify the depth of subsoil beneath the seabed and any land within and the height of the airspace above that shall be included along with the waters and seabed of the section. There are also provisions so that an area cannot cease to be part of the Marine Park except in accordance with a resolution passed by both houses of the Commonwealth Parliament. S 32 and 33 cover the procedures for preparation and acceptance of zoning plans. The Authority is required to prepare a zoning plan for a section of the Marine Park as soon as practicable after proclamation. A zoning plan may consist of one or more zones.

The essence of the underlying approach of conservation and multiple use is spelled out in S 32 (7), which defines objects to which the Authority must have regard in preparation of a zoning plan:

(a) The conservation of the Great Barrier Reef.
(b) The regulation of the use of the Marine Park so as to protect the Great Barrier Reef while allowing the reasonable use of the Great Barrier Reef Region
(c) The regulation of activities that exploit the resources of the Great Barrier Reef Region so as to minimize the effect of those activities on the Great Barrier Reef
(d) The reservation of some areas of the Great Barrier Reef for its appreciation and enjoyment by the public.
(e) The preservation of some areas of the Great Barrier Reef in its natural state undisturbed by man except for the purposes of scientific research

Sections 32 and 33 of the Act make detailed provision for substantial public participation and for ministerial and parliamentary involvement in the proCesses of developing and approving plans for the Great Barrier Reef Marine Park. Thus, sections 32 (4) and (8) provide for statutory periods, both of not less than a month, for receipt of public comment during the course of development of a zoning plan. S 32 (10) requires the Authority to submit to the minister any representations received from the public in response to the opportunity to comment on the draft zoning plan and its comments on those representations. S 33 provides for the minister to lay the plan before both Houses of the Parliament.

The Act thus provides for the establishment of a multiple-use Marine Park with objectives that include providing for sustainable exploitation of resources as well as preservation and recreation and scientific research. The only activity or use of the Great Barrier Reef Marine Park that is explicitly forbidden under the Act is the recovery of minerals. Section 38 states that no operations for the recovery of minerals shall be carried on in the Marine Park except by, or with the approval of, the Authority for the purpose of research and investigations relevant to the establishment, care and development of the Marine Park or for scientific research.

In S 66 wide powers are given to the Governor-General to make regulations under the act. Of particular importance is S 66 (2) (e), which provides the power to make regulations "regulating or prohibiting acts (whether in the Marine Park or elsewhere) that may pollute water in a manner harmful to animals and plants in the Marine Park."

Another particularly significant element of the legislation is contained in S 66 (6): "Subject to sub-sections (7) and (8) and to any contrary intention appearing in a law made after the commencement of this Act, a provision of the regulations has full force and effect notwithstanding that it is inconsistent with a law of the Commonwealth made before or after the commencement of this Act." Subsections (7) and (8) state that provisions of the regulations regulating navigation in, and flying of aircraft over, the Marine Park do not have any force or effect to the extent that they are inconsistent with a law of the Commonwealth unless such provisions can be complied with without contravention of that law.

The precedence provided by S 66 (6) is critically important since it provides the means whereby priorities of environmental management can be sustained even though they run counter to shorter term interests of administrative or economic power structures. It need rarely be invoked, but its existence obliges other, traditionally powerful, interests to consider, conduct and if necessary justify their activities in the framework of environmental sustainability.

Chapter 11

The Great Barrier Reef Marine Park: Creating An Operational Basis

---

The Great Barrier Reef Marine Park Act received assent in June 1975, but the Great Barrier Reef Marine Park Authority was not established for more than a year. Late 1975 was a time of political turmoil and increasing pressure on government expenditure. By the end of the year a new government had been elected. Where practicable, the new government froze and reviewed the initiatives of its predecessor including the Great Barrier Reef Marine Park. After review it decided to proceed with the Park and approached the Queensland Government for nominations for a member of the Authority and for members of the Great Barrier Reef Consultative Committee. The Queensland Government complied and nominated one of its most senior officers to be a member of the Authority. The Authority held its first meeting in August 1976 and shortly thereafter embarked upon a series of informal meetings with groups and individuals who were known to have particular interest in the Great Barrier Reef Marine Park.

POLICY

The earliest phase of the Authority's existence was dominated by explorations at the higher levels of policy development. Officials of the Commonwealth and Queensland governments were exploring the possibilities and implications of declaration and management of the Great Barrier Reef Marine Park.

Having established the Authority, the Commonwealth government was keen to proceed and was under strong pressure from environmental groups to do so. The Queensland government was cautious regarding the new Commonwealth initiative and concerned to avoid action that would see the state's traditional roles in the Great Barrier Reef Region further eroded.

The issues raised by the Great Barrier Reef Marine Park had broader ramifications. Central to these were section 38 of the Act, which effectively precluded commercial exploration and production of minerals in the Marine Park, and section 66 (2) (e) which provided powers for the Governor-General to make regulations regulating or prohibiting activities outside the Marine Park—"that may pollute water in a manner harmful to animals and plants in the Marine Park."

In the case of section 38, there were still valid petroleum exploration leases covering almost all of the Great Barrier Reef Region, although these had been suspended in 1970 as a consequence of the moratorium that preceded the establishment of the Royal Commissions on Petroleum Exploration and Production.[1] The governments had still not responded to the recommendations of the Royal Commissions, which were published in late 1974. The Royal Commissions had been divided on the acceptability of oil drilling in the Great Barrier Reef Region with the majority recommending limited and tightly controlled exploration in a small part of the Region and the Chairman, in a minority view, opposing any drilling because the risks of accidental damage were too high.

The mining industry identified an issue of principle. It did not accept the premise that exploration and production of oil or minerals were completely incompatible with environment management and conservation. It saw acceptance of the concept that mineral exploration and production could, on environmental grounds, be totally excluded from large areas as having severe economic consequences. Such over-riding of mineral rights could erode national and international investor confidence in the long-term and highly costly business of identifying significant mineral deposits and bringing them to production.

The mining industry saw section 38 of the Great Barrier Reef Marine Park Act as an important test case that should be fought on a number of grounds. The first was that the interests of the Great Barrier Reef petroleum exploration lessees should not be overridden without addressing the recommendations of the Royal Commissions or without compensation. The second was that the industry strongly opposed the concept that environment protection and conservation could override mineral development possibilities. Following from these was the tactical proposition that before any environmentally significant areas were closed to mining they should be prospected so that the economic consequences and compensation entitlements arising from such a decision could be assessed. A major strategic position was to take the long view—although public sentiment was opposed to oil drilling on the Great Barrier Reef, public attitudes would change with time as the price of fuel rose and as other petroleum reserves in Australia were exhausted.

It was an equally important test case for conservation groups.[2] They argued that, even if a major oil field were discovered, the benefits would be short term and would be exhausted in a period of decades, but the damage to one of the world's most significant natural environments would be long term. They strongly opposed the concept that mineral exploration and production should override natural environment management. They saw that if the case for protection of the Great Barrier Reef could not be sustained in the relatively affluent nation of Australia, the likelihood of protection of other globally significant environments, particularly in developing nations, would be small. They strongly opposed the concept of exploratory drilling before declaration of the Marine Park.

The main strategy of the conservation groups was to cite the evidence of recent marine oil well blowouts in various parts of the world as demonstrating that despite claims of the excellence of oil spill and blowout prevention technologies, the history of marine oil drilling was one of operator error and equipment failure. They cited the view of the Chairman of the Royal Commissions on Exploratory and Production Drilling for Petroleum on the Great Barrier Reef that for the immediate and foreseeable future the combination of technology and operational procedures was just not adequate. Tactically they countered calls for exploration before Marine Park declaration by identifying them as a wasteful diversionary tactic. If such exploration were to be funded by the government it would be a waste of research money that could better be spent on other aspects of the Great Barrier Reef; if it were done commercially, the venturer would certainly expect to be able to capitalize on any discovery. If exploration discovered an apparently commercial oil field, there would be virtually irresistible pressures for immediate, unsafe production drilling.

These positions had been developed and refined over more than a decade. Until a decision was made it was clear that both the Commonwealth and the Queensland governments would come under increasing and diametrically opposed pressures. The conservation groups would continue to press for proclamation of the entire Great Barrier Reef Region as the Great Barrier Reef Marine Park so that mineral exploration and production would be precluded by S 38. The petroleum industry with the support from other mining interests would fight to oppose proclamation, or at least to minimize the extent, of any section of the Marine Park while S 38 applied.

There were no precedents for the scale and form of marine environment management provided by the Great Barrier Reef Marine Park Act. It had the potential to set a number of far-reaching precedents. Despite mounting pressure for declaration of the Great Barrier Reef Marine Park, both governments saw a need to explore and find solutions to the potential problems before declaration of the first section.

In the case of S 66 (2)(e), which gives the Governor-General the power to make regulations "regulating or prohibiting acts (whether in the Marine Park or elsewhere) that may pollute water in a manner harmful to animals and plants in the Marine Park", Queensland was concerned over possible erosion of sovereignty. The concern was that the Commonwealth might seek to regulate a wide range of activities in the eastern coastal catchment that were the constitutional responsibility of the state. It could be anticipated that the more the Marine Park extended into the coast, particularly in heavily used and populated areas, the greater was the likelihood of the Commonwealth Government being persuaded to use S 66 (2)(e) to control water quality and land use and other aspects of coastal zone management. This concern was increased as conservationists and scientists advocated such a course.

A further issue was that, even though the validity of the Seas and Submerged Lands Act (1973) had been upheld by the High Court of Australia in 1974, the states expected that the new conservative Commonwealth Government would legislate to return to them some or all of their earlier powers over the three-mile territorial sea. It was possible that, following such legislation the State might see a need to take related matters to court for decision. If the State were to cooperate unreservedly in the establishment of areas of the Commonwealth's Marine Park in locations that might be the subject of later legislation or litigation, it could be seen to have prejudiced its interests.

Counterbalancing the concerns of the State was the fact that since the Great Barrier Reef Region had low water mark as its boundary with Queensland islands and the mainland, the Great Barrier Reef Marine Park could not protect the intertidal environment in such areas. Early investigations had shown that many of the areas in greatest need of management were those adjacent to islands. Some of these areas, e.g., Green Island and Heron Island had earlier been protected by Queensland National Park status. A collaborative approach was needed for sound management of marine environments from above high water to the boundaries of the Great Barrier Reef Region.

Clearly there had to be some specific arrangements to cover the concerns of both governments and to provide a basis for effective implementation of the Great Barrier Reef Marine Park Act. The task of policy preparation was to develop areas of agreement between the Commonwealth and Queensland Governments that could enable effective management of the Great Barrier Reef Region and adjacent areas. Negotiations over many months resulted in detailed, mutually understood positions on a number of points:

1. The Commonwealth was committed to proceed with the implementation of the Great Barrier Reef Marine Park Act 1975 without amendment of section 38 or section 66 (2) (e)
2. The Queensland government would continue to use its legislation for the management of areas adjacent to the Great Barrier Reef Region to the fullest extent of its jurisdiction
3. Acceptance that uncoordinated management actions were possible but that they would be costly and ineffective because of considerable duplication and would be a source of continuing legal challenge and friction over boundary definitions
4. The Great Barrier Reef Marine Park Act provided and the Commonwealth was prepared to use, considerable scope for Queensland government agencies to conduct much of the management of the Great Barrier Reef Marine Park as delegates of the Great Barrier Reef Marine Park Authority

The preparatory negotiations by officials cleared the way for a meeting between the Prime Minister of the Commonwealth of Australia and the Premier of the State of Queensland at the town of Emerald. The Emerald Agreement which emerged from that meeting contained the following elements:

1. The Great Barrier Reef Marine Park Act would be implemented with no substantial amendment
2. Queensland would introduce legislation that complemented the Great Barrier Reef Marine Park Act in order to provide consistency of management of adjacent areas
3. A Great Barrier Reef Ministerial Council would be established to coordinate policy between the two governments
4. Day-to-day management of the Great Barrier Reef Marine Park would be carried out by agencies of the Queensland Government.

DEVELOPING TECHNICAL PROCEDURES

While the policy negotiations were in progress, the Authority and its staff were exploring the technical options for implementation of the Great Barrier Reef Marine Park Act. There were no precedents for the scale and form of marine management provided by the Act. In the international arena the meetings that led to the publication of the World Conservation Strategy in 1982[3] and of the associated broader concepts of ecosystem management were just underway.

There were three interlocked areas of activity:

- Assembling information on the ecosystem, resources and use of the Great Barrier Reef Region
- Developing approaches to planning and management
- Determining the number, extent, and sequence of sections of the Great Barrier Reef Marine Park that should be recommended for proclamation

ASSEMBLING INFORMATION

The state and extent of scientific knowledge of the Great Barrier Reef and of reef ecosystems generally were reasonably well known to the Authority which had good contacts with active researchers. The Proceedings of the International Coral Reef Symposia of 1973 (held on the Great Barrier Reef) and of 1977 (held in Miami) provided solid reviews of current investigations.[4]

There was however, little information on the nature or the social or economic importance of the various human uses or impacts that the Great Barrier Reef Marine Park was supposed to manage. Two major research projects were commissioned. A survey of users from all of the coastal communities that used the Reef for recreation, or commerce was conducted by interview.[5] A bibliography of the Great Barrier Reef was commissioned to assemble the scientific and the not-so-accessible "gray" literature of government reports, press reports and nonscientific specialist articles.[6]

The lack of information on resources and on past, present and likely future use was a matter for concern in view of the requirement in S 32(7) of the Great Barrier Reef Marine Park Act to plan for "reasonable use." A major task of the early years of the Authority was thus to develop a basis for assessing what constituted reasonable use. The term "reasonable" is defined by the Oxford English Dictionary as "in accordance with reason, not absurd, not greatly less or more than might be expected; tolerable, fair". The American Heritage Dictionary (1969) has "governed by or in accordance with reason or sound thinking", "within the bounds of common sense," "not excessive or extreme".

The early studies of resources and the results obtained by Domm[7] suggested that much of the information needed for developing the understanding of the uses of the Great Barrier Reef and the extent to which they might be reasonable could only come from consulting users.

The Great Barrier Reef Marine Park Act requires public participation at two phases in the development of zoning plans but there was a need to explore the possible form and functions of public participation programs that might be carried out within the meaning of the Act. A study contract was let to a group of planning consultants with extensive experience in the United States and Australia in public participation programs for the land use and local urban and suburban amenity planning.[8] The study developed and recommended a model based on formalized public hearings and the establishement of local representative committees, which would act as facilitators and provide an interface for communication between the Authority and the public.

The Authority and staff viewed this approach with some caution. In an area as large and with a population as small as the coast adjacent to the Great Barrier Reef Region, the matter of establishing local committees truly representative of all interest groups would be difficult. An element of the difficulty would be that working commercial fishermen tend to spend most of their time at sea; many are uneasy in and distrustful of meetings. Few are prepared to discuss and debate together, much less in public, the location and relative productivity of fishing grounds. Recreational fishermen would present similar problems because membership of fishing clubs and associations was low—apparently attending meetings to discuss fishing was not seen as an important part of the activity. Subsequently figures obtained for New South Wales[9] showed that less than 2% of anglers belonged to clubs and that few of those played any part in the meeting and policy agendas of the clubs.

Early staff reviews revealed that there were serious shortcomings in the map coverage of the Great Barrier Reef region. There were various estimates of the number of reefs comprising the Great Barrier Reef but no authoritative count. Most mapping and survey had been conducted for the purposes of identification and charting of safe shipping lanes. As a consequence, areas distant from shipping lanes were poorly mapped if at all and a number of charts current at the time had extensive areas labeled "unsurveyed maybe shoal". Other sources included aerial photographs including a series of the southern half of the great Barrier Reef taken in 1966–67 for petroleum exploration mapping and maps prepared by approximate rectification of those photographs. There were many discrepancies between the various sources. A contract was therefore let with the Geography Department of James Cook University of North Queensland to develop a geomorphological classification, a master list, and map at a scale of 1:250,000 of all the reefs and shoals of the Great Barrier Reef Region.[10] At the same time a long-term program of survey was started with the Australian Survey Office with the objective of developing a series of detailed maps of samples of reefs covering all types and geographic areas of the Great Barrier Reef. For the longer term the Authority was seeking means of harnessing the obvious potential of satellite remotesensed data for the development of map-type products.[11]

The activity of the preparatory phase can be indicated from a listing of the research projects commissioned and papers prepared during the period 1976–82.[12] This is summarized in Figure 11.1.

APPROACHES TO PLANNING

The concept of zoning contained in the Act invited comparison with the practices of zoning that had been developed in town and land use planning.[13]

There were some obvious differences since the concept of the sea and marine resources as common property is fundamental to jurisdiction and management in the marine environment. The higher degree of linkage between marine environments suggested that the concept of dividing an area into separate subunits and managing them independently of each other has less applicability in marine environments than it does on land. Nevertheless there was an obvious need to investigate the extent to which principles and practices of town and land use planning could be applied or adapted for the planning of the Great Barrier Reef Marine Park. The initial study was conducted by a senior planning officer of the Queensland Co-ordinator-General's Department. The report of this study recommended a substantial study of issues that would be raised by preparation

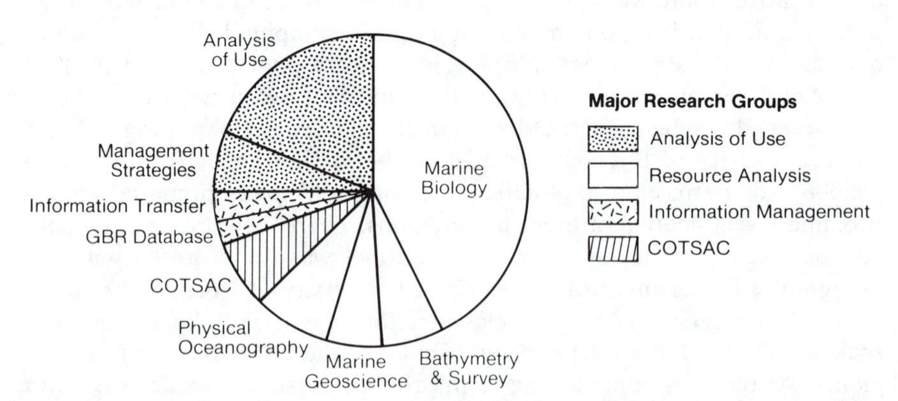

Figure 11.1 Diagram illustrating the proportion of research expenditure on major topic areas in the period 1976/86. Total expenditure approx AUS$3.28 million. COTSAC refers to projects funded under the Crown of Thorns Starfish Advisory Committee of the Authority.

of a zoning plan for a specified study area. The proposal was that the study be conducted by a consultant working with a study group comprising officers of the Great Barrier Reef Marine Park Authority and selected experts including members of the Great Barrier Reef Consultative Committee and officers of relevant Queensland Government departments and agencies. The Authority accepted this recommendation and let a contract for a major zoning strategy study.

The next issue was to select a part of the Great Barrier Reef Region as a study site for the investigation. A daunting problem was the lack of information for any but a small number of points on the Great Barrier Reef. After initial overview of the whole Region, three options were considered more closely. The first two concerned areas where there was a body of scientific information, some knowledge of usage patterns, and the likelihood that expanding use would require early management. In this category, the reefs of the Capricorn Ridge in the south of the Great Barrier Reef Region offered an accessible area in which management problems had been identified.[14] The reefs off the coast in the Cairns region (approximately 15ºS to 18ºS) offered another area with increasing use and some identified management problems.[15] The third option was based on a different approach, that of selecting a remote area, little known, little used and therefore presumably still in pristine condition and amenable to management before problems developed. For this approach, the far north of the Great Barrier Reef Region appeared to be a suitable site.

In the event, the Authority decided to concentrate initially on the reefs, shoals and seabed of the Capricorn Ridge because more was known about this than any other area of the Great Barrier Reef Region. The Capricorn Ridge had three attractions as a prototype study area. The first was its diversity in a discrete area away from the immediate influence of the coast and separated from the main mass of reefs of the Great Barrier Reef. There are 27 reefs and 5 shoals on the Capricorn Ridge and 14 of those reefs carry islands. The second, as a consequence of the research stations at Heron Island and One Tree Island the scientific literature was more extensive than for any other part of the Great Barrier Reef Region. The third was that it was a heavily used area with some partially described potential management problems of resource competition and of use-related deterioration. This decision was strongly influenced by the scarcity of information on the Cairns and far northern areas.

Although the Authority decided to concentrate its greatest effort on the Capricorn reefs, it also explored the option of the remote little used area. This was done by holding an expert workshop in April 1978 to discuss approaches to identifying and obtaining the most critical information requirements for planning and management of the most northerly quarter of the Great Barrier Reef Region.

The initial studies of the Capricorn/Bunker area had a number of parallel components. To establish the scientific background the Great Barrier Reef Committee prepared a review of the literature relating to the Capricorn Ridge specifically and to the southern Great Barrier Reef generally[16].

In June 1977 the Authority placed a public advertisement inviting public comment on the suitability of an area including the Capricorn Ridge for proclamation as the first section of the Great Barrier Reef Marine Park. The response was small and generally unsurprising. Conservation groups welcomed the proposal but said that it should occur in the context of proclamation of the entire Region as the Marine Park. Representatives of the oil industry urged that the area not be proclaimed in its entirety because it had not been fully prospected.

Design of the terms of reference for the zoning strategy study posed some problems since it would involve most of the operations specified in the Act for developing a zoning plan. It was important that the study could not be construed as developing a zoning plan and thus pre-empting proclamation by the Governor-General or the procedures set out in S 32 (7) of the Act. The solution lay in designing the study to produce three zoning plan solutions, each based on an extreme interpretation of the provisions of S 32 (7) from the perspective of a fanatical member of one of the three major use or interest categories; fishing/collecting; recreation /tourism and conservation/science. Early explorations indicated a wide range of possible interpretations of the objects defined in S 32 (7) of the Act. In particular, comments indicated that the terms "reasonable use" in S 32 (7) (b) and "appreciation and enjoyment by the public" in S 32 (7) (d) were capable of many interpretations.

The study was carried out by a zoning strategy study group comprising officers of the Authority, of the Queensland Government, members of the Great Barrier Reef Consultative Committee, and a commercial consultant. Its report[17] provided many insights into the issues and drew together much apparently disparate information. It provided a sound information base for the later development of the zoning plan for the Capricornia section, but its greatest immediate value lay in the opportunity that the study provided for those who were to be involved in the planning process to debate the issues and develop mechanisms for finding possible solutions to problems in a situation that was not pressured by a program with a high political profile and a tight deadline.

DETERMINING THE EXTENT OF SECTIONS

In section 31 the Act clearly provided for the park to be proclaimed in sections but gave no guidance regarding the extent of sections or the ultimate extent of the Marine Park. The options ranged from immediate or gradual proclamation of the entire Great Barrier Reef Region to selection of a number of separate sections to develop a Marine Park consisting of representative examples of the ecosystems expressed in the Great Barrier Reef.

In section 32(7) the Act provided for multiple and reasonable use of the Great Barrier Reef and of the resources of the Great Barrier Reef Region to be considered as objects in developing zoning plans for the Marine Park. With the exception of the prohibition in S 38 of operations for the recovery of minerals, other than for purposes of approved research, no use of the Great Barrier Reef or of the resources of the Great Barrier Reef Region was automatically precluded from the Marine Park.

The decision regarding the eventual size of the Marine Park would require development of criteria to guide decisions on areas that should be included or excluded. Objective criteria for inclusion would require a high degree of understanding of the extent of the ecosystems of the Great Barrier Reef, of their interdependence, and of their vulnerability to impact from human use of adjacent areas. Objective criteria for exclusion would require a high degree of understanding of the extent, interaction and consequent impacts of existing and potential future uses upon the ecosystems of the Great Barrier Reef.

Information on the nature and functioning of reef ecosystems was patchy and much of it related to studies of conditions and processes at small scales on parts of reefs. Nevertheless, it was clear that transport of nutrients, food, and larval recruits occurred over long distances through the movement of water currents. Thus, apparently separated areas could be closely linked. In the absence of more precise information the decision that forfeited least options for conservation and management in accordance with the act was to include large areas.

There was widely expressed concern at the actual or potential impact of most of the activities that directly exploited the resources of the reef or that threatened them indirectly through pollution. Despite this there was virtually no information on the extent or the impacts of commercial or recreational fishing, collecting, tourism, reef walking, or diving. There was no reason to consider any of those activities as inherently unreasonable at the levels occurring in late 1977. There was therefore no reason to suppose that they should not continue subject to management within most of the Great Barrier Reef Marine Park.

These considerations led to the conclusion that sections should provide the capacity to manage or regulate impacts within the Marine Park, and to buffer the more highly protected zones from impacts originating outside the Marine Park. This suggested that the Marine Park should probably be very large in order to ensure that all ecosystems of the Great Barrier Reef were included. However, even though the practical considerations of planning and implementing management for conservation on such a scale were unexplored, it was clear that there were many potential conflicts. Some were likely to arise because of the jurisdictional sensitivity of any areas in which the Great Barrier Reef Marine Park might come to the low water mark on the mainland and have consequent implications for the use and management of adjacent areas of Queensland. Others would arise simply because the Great Barrier Reef Marine Park Act was a new and broadly based piece of multiple-use management legislation which would now apply to activities that had previously been unregulated or subject to special interest legislation.

The first Section to be proclaimed was the Capricornia Section which did not extend to the mainland coast. It encompassed the Capricorn /Bunker group which was the subject of the zoning strategy study.[18.] This enabled the technical details of zoning and management preparation to be followed through without the immediate and major additional complication of mainland issues.

Chapter 12

The Great Barrier Reef Marine Park: The Planning Procedures Adopted By the Authority

---

The development of a zoning plan is a complex process. It takes about two years from initial preparation until the approval by the Parliament. It involves the coordination of many Authority and Queensland officers, and extensive input from other organizations and the public. Public participation and consultation are centrally important but time-consuming components that form the core of the zoning process.

THE ROLE OF PUBLIC PARTICIPATION IN PLANNING

At the start the Authority faced a critical decision concerning the role and conduct of the public participation programs specified by the Act. One option was to follow more or less conventional land use planning practice and publish modest formal notices of plan development and of availability of the prepared plan for comment. Under this option the Authority would conduct reactive public participation programs that did the minimum necessary to meet the requirement under the Act for it to present its expert work for scrutiny and possible objection by groups or members of the public conversant with such procedures.

The alternative, adopted by the Authority, was to develop proactive public participation programs as integral parts of its processes of information gathering, planning and education. This approach seeks to maximize two-way information flow.

As a means of gathering information, proactive public participation draws upon the profound knowledge that some users are likely to have about the area under consideration. The knowledge and understanding developed by fishermen, tourist operators, or beachcombers who spend most of their time on or near the water can bring new perspectives on natural resources and on changes that have occurred over a number of years. The information may lack some of the precision of professionally designed studies, but it complements and extends the slender information base of formal investigations.

As a means of disseminating information, public participation can address the fact that those who use an area will be most directly affected, interested in, and suspicious of any proposed management plan. If they are informed and consulted in management planning and their concerns understood and taken into account, they are more likely to understand and support the objectives of management.

Considerable effort is put into management of the public participation programs in order to make them efficient and nonbureaucratic. There appears to be a basic apathy and cynicism expressed in such terms as "Why bother? You're only asking because the law says you have to. Nothing that I say will make any difference—in any case you probably won't even read it".

In the development of a program the focus of effort is to make public participation materials attractive and clearly expressed. The core of each program is a colorful mail-back brochure designed to project a "user-friendly" approach, to provide basic information on the planning program and to encourage users to respond to key questions. The brochure encourages people to obtain a booklet of more detailed information and to contact Marine Park staff with requests for further information or for meetings with planning team staff.

Once the program has started a telephone is permanently manned during working hours by staff trained to ensure that:

- Questions are answered immediately or, with explanation to the caller, referred to a named expert officer
- Requests for the materials are filled promptly
- Requests for meetings with Authority staff are logged and arrangements made for them to be fulfilled

The terminology of the Act calls for the public to make representations to the Authority and for the Authority to make submissions to the minister. This terminology is followed in the public programs because the semantics contain an important issue. Members of the public are invited as equals to represent their position and information to the Authority. The Authority is not a superior body in the planning process; it is simply the appointed body charged with assembling and analysing the information.

The Authority submits recommendations and proposals to its minister.

Feedback is important. When a representation is received, it is logged and a form letter of acknowledgement is sent, if possible within three working days. The name and address of the respondent is entered in a mailing list so that each can subsequently receive a letter and a copy of the materials developed as a result of the public participation program. At the conclusion of the process all participants receive a copy of the plan and those who commented on the draft plan receive a letter that briefly presents the Authority's response to the issues raised.

The form and style of representations vary. The majority use the mail-back brochure and respond to the questions. A few consist of detailed technical analyses and arguments that may cover many pages. Each representation is independently summarized and coded for computer entry and indexing by two officers. Differences are resolved with the project officer and other planning team members. The factors entered include the activities mentioned and a summary of the nature of the activities, areas or locations that are referred to specifically. Computer summaries and indexes permit a variety of reports to be produced. Reports refer to individual representations by index number so that it is possible and usual for a planning team member to retrieve and read firsthand all comments about a particular location or activity.

Robinson[1] noted that the interests of local and nonlocal users of environmentally significant areas tend to differ. Those living in or near such areas are more likely to participate in the planning process and are more likely to focus on the potential for harvesting and extractive activities often with very site-specific information. Those who come from farther afield are less likely to participate because they have less site-specific information but, if they do, they are more likely to focus generally on leisure, recreation, environment appreciation and protection for the future. He urged planners to be cautious in seeking a balance of local and national interests.

Reports are produced that cite the number and point of origin, by postcode, of representations received. These give an important indication of the geographic extent of effective contact and involvement in the public participation process.

The results of the public participation programs are treated cautiously since the programs are not designed to obtain statistically representative samples of the opinion of the public generally or of subsets of the public. The structure and promotion of the public participation programs are designed specifically to obtain comment from interested parties. There are no safeguards or identity qualifications to prevent an individual or group sending large numbers of representations.

Considerable effort is expended during the program to avoid participants or staff coming to regard representations as votes. This reflects the underlying position that a piece of information or an argument relating to achievement or interpretation of the objects of the Marine Park is as carefully evaluated if it is made in one representation as in hundreds.

Despite this, in an operation designed to develop consensus, some guidelines are needed to assess the breadth of opinion. These permit some cautious consideration of the claimed and apparent representativeness and the effort behind representations. This is done qualitatively in order to avoid quantitative rules that could rapidly be abused to create a quasi-voting system.

The guidelines are:

- Each representation receives a unique number
- On the basis of content and any address or heading on the notepaper, a note is made whether the representation comes from an organized professional or voluntary group or company. In the public participation materials such groups are asked to indicate the extent of their membership or shareholding and whether the representation was considered and adopted by a general meeting. Thus where possible, the claimed membership and representativity of the organisation are noted. The number of signatures on a letter from an informal group or association is noted
- Duplicate or form letters are treated by assigning a number after a decimal point to reflect the number of separate representations presenting the same material

THE PLANNING PROGRAM

There are five stages in the development of a zoning plan for a section of the GBRMP:

1. Initial information gathering and preparation: Staff and consultants assemble and review information on the nature and use of the section and develop materials for public participation
2. Public participation—notice of intent to prepare a plan: Members of the Authority and staff seek public comment on the accuracy and adequacy of review materials and suggestions for content of the proposed zoning plan
3. Preparation of draft plan: Preparation and Authority adoption of a draft zoning plan and materials explaining the plan for public participation
4. Public participation—Review of draft plan: Members of the Author-

ity and staff seek public comment on the published draft plan and explanatory materials
5. Plan Finalization: The Authority adopts a revised plan that takes account of comments and information received in response to the published draft plan, and submits this to the minister

A planning team is established with a nominated project officer who is responsible to a senior officer for the conduct of all tasks of the specific planning program until the plan has been accepted by the Parliament. The planning team comprises members of the technical sections of the Authority's staff and, usually, a representative of the Queensland managing agency. The planning team meets to discuss strategies, to consider progress reports and future work and allocate specific tasks, with completion deadlines, to nominated officers. It is usual for the Chairman of the Authority to attend critical planning team meetings at which major potentially controversial zoning issues are debated. Reports arising from the work of the planning team are considered by the Great Barrier Reef Marine Park Authority, which makes decisions in relation to recommendations and alternative approaches identified by the team.

The planning team comprises officers from all of the technical sections of the Authority's staff. Broadly, tasks relating to the synthesis of published scientific and technical information and internal reports and to commissioning of specific research or investigation projects fall to officers of the Research and Monitoring section. Tasks relating to feedback from existing day-to-day management and the logistic requirements of future management are carried out by officers of the Park Management section in consultation with the Queensland managing agency. Tasks relating to the design and production of materials for public participation programs and documents are carried out by officers of the Information and Education section. The overall conduct of the program, the preparation of reports of public participation and the development of plans and associated documents are the task of the Planning section. In practice the arrangements involve considerable project management to provide the flexibility, task sharing and swapping needed to ensure that deadlines are met for the plan and also for the range of projects underway in other major programs at the same time.

INITIAL INFORMATION ASSEMBLY

The initial task of the planning team is to assemble and review available information on the resources and use of the area to be planned and, if the area is already under management, on the experience, effectiveness, and performance of management. From this initial review specific investigations may be identified as necessary to provide important information within the available time frame for the current planning operation. A review document is then developed for the planning team.

The review document provides the basis for development of public participation materials. The most widely distributed of these is a descriptive brochure that describes the purpose of the program and the process of plan development and invites interested readers to contact the Authority for further information. The brochure incorporates a map, a questionnaire and a paid mail-back panel to make it as easy as possible for a respondent to make a representation. The most substantial published document is a jargon-free summary of approximately 50 pages (e.g., GBRMPA)[2] which is mailed to individuals and groups on a mailing list and sent to those who request further information after reading the brochure. This document seeks to draw out the issues that must be faced in developing the plan and a major part consists of maps illustrating the distribution of resources and usage patterns.

Preparations for public participation require the development of a theme for the brochure and advertising materials for press, television, and radio. The aim is to develop materials that will attract public attention and be sufficiently different from previous planning programs to avoid confusion or saturation.

The final preparation element involves arrangements to publicize and distribute the brochure and other materials. Summaries and brochures are mailed to a large number of groups and individuals on contact lists maintained by the Authority and the Queensland managing agencies. Otherwise the primary means of distribution is through small promotional display panels with a supply of brochures. These are set up on shop counters and in offices of organizations that have interests relating to the area being planned and are prepared to replenish the brochures in the display bundle provided. An important preliminary logistic task is the development of itineraries and agreements for deployment of the countertop displays and for contact with the press and electronic media for publicity at the launch of the program. Once the program has been launched this task evolves into organization of a schedule of meetings to accommodate requests by interested groups for meetings to discuss the planning program with Authority members or staff.

PUBLIC PARTICIPATION—NOTICE OF INTENT TO PREPARE A PLAN

The primary function of this phase is to alert reef users and those interested in the Great Barrier Reef that a zoning plan is to be prepared. It is used to seek comment and correction of maps and other information concerning distribution and use of resources and to solicit opinions on appropriate provisions to be included in the zoning plan. Where the planning program involves review of an existing, implemented zoning plan this phase in an opportunity to test the results of research into user reaction.

In the cycle of development of initial zoning plans for the Marine Park, this phase had the difficulty that, in the absence of specific zoning proposals, most respondents had generally supportive views on the need for management but little, if any specific information to add and few specific proposals. Whereas the Authority and its staff accepted that it was unreasonable at this stage to expect specific carefully argued proposals, comments at meetings and in some representations indicated that some felt frustrated, resentful, or cynical about a government public participation program that invited specific proposals that few if any, were in a position to make. After initial encounters with this reaction officers attending meetings and responding to telephone calls and correspondence stressed the information role of this phase of the program and the importance of respondents commenting on the specific proposals in the subsequent public participation program to review the draft plan.

PREPARATION OF DRAFT ZONING PLAN

The draft zoning plan is developed according to the objects established in S 32 (7) of the Act. In this the planning team works to a series of guidelines, expressed with the preamble "as far as practicable", which help to develop a contemporary interpretation of reasonable use and to ensure that multiple use objectives are properly considered.

Guidelines for Preparation of the Mackay/Capricorn Zoning Plan

General, Legislative and Management Requirements

1. The zoning plan should be as simple as possible.
2. As far as practicable, the plan should minimize the regulation of, and interference in, human activities, consistent with meeting the goal of

providing for protection, wise use, appreciation, and enjoyment of the Great Barrier Reef in perpetuity.

3. As far as practicable, the plan should maintain consistency with existing zoning plans in terms of zone types and provisions
4. As far as practicable, the plan should maintain consistency with plans drawn up under Queensland Marine Parks legislation
5. As far as practicable, the pattern of zones within sections should avoid any sudden transition from highly protected areas to areas of relatively little protection. The concept of buffering should be applied
6. As far as practicable, unless levels of localized activity suggest otherwise, single zonings should surround areas with a discrete geographic description, e.g., an island or reef
7. Zone boundary widths should be consistent around reefs and islands and where possible should be described by geographical features (based on line of sight to aid identification in the field)

Specific Requirements

Shipping

1. The plan should provide for the movement of shipping along recognized or proposed routes
2. The plan must not impede the access of international, interstate, or intrastate shipping to shipping routes or routes into existing ports on the coast of Queensland. Nor should it impede access to potential ports

Defense Areas

1. The plan must recognise the requirements of the Department of Defense, particularly with regard to gazetted defense areas

Conservation of Significant Habitat

1. As far as practicable, areas of world, regional, or local significance for wildlife conservation (involving, e.g., dugong, whales, turtles, crocodiles) should be given appropriate protective zoning.
2. As far as practicable where significant breeding or nursery sites can be identified, particularly for species subjected to harvesting, these should be provided with appropriate seasonal closure, Marine National Park, or Preservation zoning
3. As far as practicable, representative samples of characteristic habitat types should be included in either Marine National Park "B" or Preservation zones
4. As far as practicable, in reefal areas protective zoning should be ap-

plied to reef/shoal complexes (i.e., to incorporate a wide range of habitat types within one unit)

5. Reefs and other areas adjacent to coastal settlements and/or popular departure points are often the focus of fishing and related activities. As far as possible a group of Replenishment Areas (closed for set periods to enable fish and other exploited resources to regenerate) should be declared within the same general area

National Parks, Reserves and Historic Shipwrecks

1. As far as practicable zoning of reefs and waters adjacent to existing national parks, fisheries reserves, and historic shipwrecks should complement the objectives of those reserves

Anchorages

1. As far as practicable, major anchorage sites should be in General Use zones to allow most of the activities associated with overnight or longer anchoring of vessels to continue. Where an anchorage is zoned in a manner that restricts those activities, as far as practicable the opportunity to carry out those activities should be provided at an adjacent anchorage
2. As far as practicable, the zoning for anchorages should not result in the multiple zoning of a single island/reef unit simply because an anchorage is present
3. As far as practicable, the plan should retain access for small boats to important all-weather anchorages. However, access to all zones during emergency conditions must be allowed for

Scientific Research

1. Provision should be made for the conduct of scientific research throughout the Section. However, areas should be zoned exclusively for scientific research only where existing and probable future research programs indicate that those areas are likely to be used for that purpose on a frequent and regular basis. In other cases declaration of areas for special management for scientific purposes should meet the needs of the scientific community
2. As far as practicable, zoning should complement current spear-fishing restrictions as set out in the Queensland fisheries regulations

Commercial and Recreational Activities

1. As a general rule:

   - areas of significance for nonextractive activities should as far as practicable be given Marine National Park zoning
   - areas of significance for reasonable extractive activities should as far as practicable be given General Use zoning

2. When a reef or area is zoned in a way that excludes a particular activity, provisions should be made in as many cases as possible for access to alternative areas

It is generally the case that a majority of zone allocation decisions flow logically from the guidelines. There are a few "toss-up" allocations where one of several, apparently similar reefs could logically be allocated to a restrictive zone. There are some sites, usually those near islands or most accessible from harbors or boat launching ramps where there are clear conflicts of use, the resolution of which will inevitably please one party but displease another.

The planning team develops a draft plan and, if necessary, alternative options for specific problem sites. These are considered by the Authority which adopts a plan for release for the second phase of public participation. The Authority also adopts a report containing information, updated as necessary since the initial public participation program, explaining the basis for zoning, and presenting a brief summary of specific reasons for zoning of any reefs or areas allocated to zones more restrictive than "General Use". These documents are used to develop second phase public participation materials that follow the style and promotional design theme adopted for the initial phase. These are a 50 page summary and a brochure containing a zoning map, a summary of zoning provisions, a list of questions concerning information of interest to the Authority, and a mailback panel for easy response.

PUBLIC PARTICIPATION AND REVIEW OF DRAFT ZONING PLAN

The program is conducted similarly to the initial phase. An information summary and brochure are widely distributed by mail using a list expanded to include all those who responded in the initial public participation program. Countertop displays make the draft proposals widely available. Meetings are arranged in response to user requests. This phase is usually easier to conduct since users find it much easier to evaluate and react to specific proposals. Material and presentations emphasize that the proposal is not final but is a draft published for the specific purpose of inviting public comment. Respondents who wish to object are invited to specify their objections, to propose alternative solutions, and to support their arguments with factual evidence where possible. Those who support all or specific parts of the plan are asked to say so in representations because in the absence of that information it is possible in revision to modify a plan to meet an objection by one user group and unwittingly overturn a solution regarded as good by another.

Representations are summarized, coded, and entered for computer indexing and analysis as they are received. Progress reports on the analysis are produced as required during the program. The aim is to produce a detailed analysis within two weeks of the last reasonable date for receipt of representations posted on the final day of the program.

PLAN FINALIZATION

The planning team meets after the report on the analysis of representations has been completed to consider the issues raised in the public participation program and to discuss and evaluate possible changes to the published draft plan. These are then considered at a meeting of the Authority and a listing of changes that the Authority is prepared to consider further is developed for discussion with the Great Barrier Reef Consultative Committee and with committees of officials of the Commonwealth and Queensland governments.

The content of the final plan is subsequently decided by the Authority. Then, after the completion of precise cartography and written boundary definitions for all zones, checking by the Commonwealth's legal officers, and final proofing of the plan and zoning maps, the plan is printed. A copy of the printed plan is then formally adopted by the Authority and, as required by the Act, submitted to the minister with a report on representations received in response to the draft plan. The minister is obliged to accept or reject the plan within 15 parliamentary sitting days of its submission by the Authority. Once the zoning plan has been accepted, the minister is obliged to arrange for it to be tabled in the Parliament within 15 sitting days of the date of acceptance. Once tabled either House of the Parliament may within 15 sitting days of tabling pass a resolution disallowing the plan. If this does not happen, the minister announces the date upon which the plan is to come into effect.

Chapter 13

Zoning Plans As A Multiple Management Approach

Three major overlapping groups use, or have interest in, the ecosystems of the Great Barrier Reef. The longest established is that of fishing and collecting or harvesting the renewable resources. The next is recreation, recently augmented by tourism. The third, which may be summarized under the general heading of conservation, encompasses scientific and other philosophical positions on the need to maintain the ecosystems of the Great Barrier Reef. It includes the vicarious user who may see books, magazine articles, films, or television coverage of the reef, is unlikely to visit, but accepts national and international responsibility for the continued well-being of an ecosystem of great significance.

A zoning plan has to balance the needs and aspirations of the three main user groups. The balance varies depending on geographic and social factors including the accessibility of reef areas, the history or tradition of resource use and the sustainability of options for economic activity based on resources of the Marine Park. Within the requirement to provide for the conservation of the reef, the development of a plan involves a series of decisions on the use of reefs, parts of reefs and interreefal areas. Each area may be allocated for fishing/collecting, recreation/ tourism, or conservation/preservation. The sum of the decisions should produce a reasonable balance between the three groups. The situation is complicated by the fact that although the demands and impacts of the groups may be very different, the groups overlap. Thus a fisher is increasingly likely to support conservation of important areas. A conservationist is likely to be a tourist, and many recreationists like to catch fish.

The requirements for fishing and collecting are that opportunities should allow for long-term sustainable access and use of exploitable species. The protection of nursery areas is generally accepted as a necessary management measure. Within the available fishing areas some techniques may clash with the access or amenity of others. Prawn trawling is often cited as affecting the habitat of juvenile fish. Seine netting takes fish otherwise available for line fishing.

Different groups may have different criteria for success as they compete for or share access to the stocks. Commercial fishing is based on the logic of the value of fish or invertebrates in the market. For a successful commercial fishery the number of participants should be no more than that whose combined sustainable catch rate yields enough product to pay all costs of catching, processing and marketing including a reasonable income or return on capital for all involved. Recreational fishing is based on the logic of the amount of money the fisher is prepared to pay for the experience of catching fish. Many studies, e.g., Gartside[1], have shown that motivation of recreational fishing is complex. Size of fish is frequently more important than number or overall mass caught in a session. The actual catching of fish, although declared as a primary objective, may be a subordinate factor to the main purpose of recreation in a natural area away from pressures of work or domestic environments. For a successful recreational fishery, the number of participants should be no greater than can have a reasonable expectation of catching a few reasonably sized fish on most fishing trips.

The primary requirement for conservation/preservation is that the combined effects of all human activities and natural catastrophes should not exceed the sustainable capacity of the environment. The secondary requirement is that there should be some areas where the public may have access to substantially undisturbed areas free from fishing and collecting in order to appreciate the nature of the environment. These may be achieved by overall control of the extent of activities, by creating some public access/no fishing or collecting zones, and by setting aside some areas to be undisturbed and thus act as reference sites, refuges or functional genetic reserves.

The reasonable expectations of conservation/preservation and fishing collection can be met by a gradation of zones from public access with no restriction on otherwise legal fishing, though restrictions based on permitted equipment, to public access with no fishing, to no public or fishing access.

ZONES

The basis of Great Barrier Reef Marine Park zoning plans is spatial control. The provisions of the Act require that a zoning plan establish the purposes for which each area of the Marine Park may be used or entered.

A use or purpose of entry may be "of right"—that is any person may undertake that use or purpose of entry subject to any condition specified in the plan. Such uses are also subject to regulations under other laws provided that compliance with those regulations is not inconsistent with the provisions of the zoning plan.

A use or purpose of entry may be allowed only after prior notification of the Authority or its delegate.

A use or purpose of entry may be allowed only with a permit from the Authority or its delegate.

A use or purpose of entry not specified as of right, after notification or by permit is not allowed unless the proponent can establish grounds to receive a permit under the category of a use "consistent with the objectives of the zone".

SPECIAL MANAGEMENT AREAS

In addition to zones, the zoning plan makes provision for areas in which different management provisions may apply to parts of zones—generally for periods less than the expected life of the zoning plan. In the Mackay Capricorn section these include:

1. Defense closure areas—which make provision for areas that may be closed under provisions of the Defence Act during military training exercises
2. Special management areas—which may be proclaimed by the Authority to cover no more than 20% of a reef or zone and may make provision for such requirements as:

   - Seasonal closure to prevent or limit use during sensitive periods in the reproductive behavior of animals of the Marine Park
   - Reef appreciation to protect from fishing or collecting areas of otherwise less protected reefs in order to maintain the amenity value of sites important for tourist or recreational reef viewing
   - Research to protect sensitive research sites from disturbance by other uses
   - Replenishment or recovery to protect areas recovering from natural or human impacts

Special management areas can be an important management tool for dealing urgently with changes to the condition or use of parts of a section of the Marine Park. They offer a means to deal with unanticipated situations, but there are limits in the extent to which they can be used. A special management area changes the purposes of use and entry for the area to which it applies. Used widely, special management areas could greatly change the pattern of use away from that identified as reasonable by the elaborate process of public participation and as a consequence approved by the Parliament.

The effect of a zoning plan upon users is illustrated by the summary Table 13.1 of allowed uses and activities for the Mackay/Capricorn section zoning plan.

THE PERMIT SYSTEM

The provisions of the Act make the zoning plan an inflexible instrument that cannot be quickly altered. This is deliberate for the purpose of ensuring that the management intentions approved by the legislature are not altered or subverted by executive government or its officials. The inflexibility also protects officials from possible political pressure to make changes. Nevertheless it limits freedom to respond to specific situations arising in the course of in daily management.

A level of flexibility is provided by a permit system that enables issues that cannot be precisely anticipated in the zoning plan to be considered on a case-by-case basis. This enables management to set conditions for, and to monitor and manage the development of, new, sensitive, or major activities. Such responsiveness is achieved at the cost of establishing a system for proper, fair, consistent, and accountable processing and assessment of permit applications, for monitoring compliance with permit conditions, and impact of permitted activities. Fairness and consistency require that a body granting permits that may have a major impact on the amenity, activities, or the intentions of users should respond to permit applications reasonably rapidly and that the amenity of other users who might be affected by the granting of a permit should be considered. Decisions of the Great Barrier Reef Marine Park Authority are subject to appeal to the Commonwealth Administrative Appeals Tribunal and the Commonwealth Administrative Decisions Judicial Review Act.

A crucial issue is the dynamics of use and impacts. What one or two people may do sustainably on a reef over many years may become unsustainable when done by 100 or 200 people, and will almost certainly become unsustainable when done by thousands of people unless management measures are introduced to minimize and control impacts. Impacts may be avoided or reduced by directing activities to areas that can sustain them[2] or by providing hardening or facilities to prevent damage, such as moorings.[3]

Table 13.1
Uses/Activites for the Mackay/Capricorn Section

Use	1	2	3	4	5	6	7	8	9	10	11	12	13	14
General Use A	YES	PMT	LIM	PMT	YES	YES	YES	YES	YES	PMT	YES	PMT	PMT	YES
General Use B	YES	PMT	LIM	PMT	YES	YES	YES	YES	YES	PMT	YES	PMT	PMT	NO
Marine National Park A	YES	PMT	NO	NO	NO	LIM	YES	LIM	YES	PMT	NO	PMT	PMT	NO
Marine National Park B	NO	PMT	NO	NO	NO	NO	YES	NO	YES	PMT	NO	PMT	NO	NO
Preservation	NO	NO	NO	NO	NO	NO	NO	NO	PMY	PMT	NO	NO	NO	NO

Uses
1 Bait netting and gathering
2 Camping
3 Collecting (recreational—not coral)
4 Collecting (commercial)
5 Commercial netting—not bait
6 Crabbing and oyster gathering
7 Diving, boating, photography
8 Line Fishing (bottom fishing, trolling, etc.)
9 Research (nonmanipulative)
10 Research (manipulative)
11 Spearfishing
12 Tourist and educational facilities and programs
13 Traditional hunting, fishing, and gathering
14 Trawling

The permit system provides for case-by-case consideration, establishment of case- or site-specific controls, and monitoring of activities that have the potential to impact upon the structure, process, or amenity of the Marine Park. A list of activities specified as requiring permits in the least restricted, General Use, zone of the Mackay/Capricorn Zoning Plan is provided below. It illustrates the range of activities considered to need case-by-case evaluation. Obviously in more restricted zones some of the listed activities are not allowed even with a permit and some of the "as of right" activities in the least restrictive zones become the subject of a permit in the more restricted zones.

The activities requiring permits in the general use zone are:

- Commercial collecting
- Manipulative research
- Construction and conduct of underwater observatories
- Provision of tourist or educational facilities
- Establishment of tourist or educational programs
- Construction or operation of mooring facilities for vessels
- Operation of aircraft
  -on the surface of the ground or water
  -at an altitude less than 500 feet except in an approved aircraft landing area
- Construction and conduct of an aircraft landing area
- Use of hovercraft
- Harbour works, beach protection works or other works
- Dumping of spoil
- Discharge of wastes from a fixed structure
- New commercial fisheries
- Traditional hunting
- Any other purpose consistent with the objectives of the zone

The categories cover activities that are likely to have some impact. For these the purpose of the permitting process is to assess the extent of the impact of the proposed activity and thus to decide whether, or under what circumstances, it is consistent with the objectives of the zone and should therefore be permitted.

The categories also include activities of unknown extent and potential impact such as collection, new fisheries or tourist programs. For these, the purpose of permitting is to monitor with the possibility of imposing controls if the activity approaches unreasonable levels.

In the case of tourist developments, such as proposals to moor or install day visitor pontoons or floating hotels, permitting triggers a thorough process of environmental impact assessment, design of specific controls, and monitoring requirements required to install and manage the development without unreasonable impact.

Within each zoning plan, for each zone there is a list of activities, which may take place with the permission of, and subject to any conditions imposed by, the Authority. The zoning plan and regulations specify the information that must be contained in a permit application; see list below.

Basic information required for permit application:

1. Name and address of person making application
2. Name assigned to section of Marine Park to be used or entered
3. Name of zone
4. Purposes for which zone is to be used or entered
5. Any prudent and feasible alternatives to the proposed use or entry
6. Proposed movement within the zone of any person proposing to use the zone
7. Location of use of, or entry into the zone including the name of any shoal, reef or island on or near which such use or entry is proposed
8. Period for which relevant permission is sought
9. Means of transport for entry into, use within and departure from the zone or area
10. Maximum number of persons
11. Such other information (if any) as the Authority may reasonably require and has requested the applicant to furnish

An application for a permission for research must also provide the following:

1. Purpose of the research
2. A brief description of the manner in which the research is to be undertaken, including:

    - Description of the sequence and location of fieldwork
    - Explanation of the experimental design and methods of analysis to be used in the research
    - Number, quantity and identity of any living or non-living matter to be taken for the purpose of research
    - Methods to be used in taking such matter

Following is a list of matters to which the Authority shall have regard in considering a permit application:

1. Object of zone
2. The need to ensure orderly and proper management of the zone
3. The likely effect of granting the permit on future options for the Marine Park

4. The conservation of the natural resources of the Marine Park
5. The nature and scale of the proposed use in relation to the existing use and amenity, and the future or desirable use and amenity of the relevant area and of nearby areas.
6. The likely effects of the proposed use on adjoining and adjacent areas and any possible effects of the proposed use on the environment and the adequacy of safeguards for the environment
7. The means of transport for entry into, use within or departure from the zone and the adequacy of provisions for aircraft or vessel mooring, landing, taking off, parking, loading or unloading
8. In relation to any structure, landing area, farming facility, vessel or works to which the proposed use relates:

   - The health and safety aspects involved, including the adequacy of construction
   - The arrangements for removal upon expiration of the permission of the structure, landing area, farming facility or vessel or any other thing that is to be built, assembled, constructed or fixed in position as a result of that use
   - The arrangements for making good any damage caused to the Marine Park by the proposed activity

There are established procedures for considering proposals for developments and related activities that may have a significant impact on the environment or on the amenity of other users. The procedures are somewhat complicated by the range of jurisdictional responsibilities within and adjacent to the Great Barrier Reef Marine Park. Depending on the location of the proposal, the zoning plans and regulations of the Great Barrier Reef Marine Park or Queensland Marine Park will require consideration in the context of permit application assessment.

The entire area within the outer boundaries of the Great Barrier Reef Region, including islands under the jurisdiction of Queensland, is a World Heritage Area.[4] This places a responsibility on the Commonwealth minister for the environment to consider any proposed activity in the context of the Environment Protection (Impact of Proposals) Act. The Authority provides the initial advice to the minister regarding proposals in the Great Barrier Reef Region. There is a graded response. If it appears that there may be a significant impact, the minister may designate the proposal as requiring assessment. For relatively minor proposals, or for proposals whose generic nature is well known such as tourist daytrip pontoons, assessment may be carried out on the basis of a Preliminary Environmental Report, a short form that addresses design or operational measures specified to avoid limit or control likely impacts. A major proposal or one with unknown or poorly predicitable impacts will generally require preparation and evaluation of a comprehensive Environment Impact Statement. Where, as is often the case, the proposal concerns an area in the territorial sea or is part of Queensland, it may also require attention under Queensland State Environment Assessment legislation. In such cases, the Authority collaborates with the Commonwealth and Queensland departments responsible for environmental matters to ensure that the necessary issues are assessed under all relevant legislation with one series of investigations or enquiries.

Claridge[5] provided a checklist for the consideration of reef related development proposals. That list, based on the U.S. Environment protection Administrative Procedures, has application for proposals in other marine environments. Typically, the environmental impact is considered in a predictive study and the results of that study are used to modify a development proposal. Permits are issued to cover construction and operation of the development.

The benefits of such a system are that it provides a means to control the development of new or changing uses while their impacts and sustainability are considered. In cases where there is potential for a reasonable use to develop to the extent that it is no longer reasonable, the permit system may prevent such an outcome. The imposition of information return conditions on permits can enable managers to monitor the development and impact of permitted activities.

The demands of such a system of permits and environmental assessment are unpredictable and usually urgent. They may be reduced by the development of clear guidelines that specify the conditions of site selection, design and operation that must be fulfilled in order to qualify for a permit. Despite such guidelines many such issues become controversial and may be subject to court action flowing from appeal provisions.

The permit procedures involve a series of steps that, undertaken systematically, ensure that issues are properly considered and that decisions are made consistently and fairly. The combined procedures provide for judgment to be applied in order to achieve an appropriate response ranging from rapid permitting of a simple uncontroversial facility such as a mooring to full consideration of a major proposal such as a floating hotel. Some simple requests, such as one for a permit for aircraft operations or tourist cruise ship reef visit can be handled in a few days but permits for major operations may take weeks or months. A permit request that involves preparation of a full Environmental Impact Statement, or public advertising in accordance with Great Barrier Reef Marine Park regulations, generally takes several months from initial application to decision.

Permits for development, installation, or operation of facilities are now usually subject to the requirement that the developer pay for the conduct of a monitoring program to assess the impacts of the activity on the environment. In the case of construction activities, the monitoring may have an immediate feedback function so that, if the construction technique causes unpredicted and excessive levels of suspended sediment, operations can be halted and another safer technique used.

Operational monitoring enables changes to be assessed with the objectivity possible from a statistically designed approach. This enables operator, manager and the observing public to understand the extent to which an activity causes detectable impacts, to make decisions on the apparent sig-nificance of the impacts and any consequential management response that is needed. Over the years as human impacts and the ability of the natural system to absorb them become better understood it can be expected that management techniques and the nature of monitoring programs will change.

A less administratively burdensome method of monitoring and setting conditions for certain nonthreatening activities is to make them notifiable. That is, a person wishing to undertake them is obliged to notify the Authority or its agent who may set conditions. Where possible without compromising the objectives of the Marine Park or of zones within plans the use of statutory notification and activity monitoring by managers or research contractors is preferred.

In revising plans and procedures the GBRMP Authority is seeking to minimize the number of activities that require permits. It is also developing clearer management guidelines on the likely conditions for permissible uses and the information requirements for assessing applications and thus to save effort, expense and time spent by permit applicants and assessors on proposals that are incomplete or likely to require major modification to reach an acceptable form. The intention is to provide a framework for the priorities of day-to-day management programs and guidance to permit applicants and to permit decision-making delegates.

As part of this approach of providing more specific guidance to managers and permit applicants, the Authority has also adopted two planning approaches that provide site specific guidance on the preferred forms of use within the strategic overview of the Zoning Plan. Area Statements provide guidance on the desired usage setting and management priorities of individual reefs in the context of the Zoning Plan. Reef Use Plans provide site-specific policy based on detailed information on the distribution and condition of resources within a reef.

AREA STATEMENTS

Zoning plans provide an immediate strategic framework for managing activities within the Marine Park. They specify the purposes of use or entry for each zone but not the level or the amenity setting. Area statements provide an additional level of strategic guidance to permit applicants and to delegates regarding the usage setting and management priorities for each reef or management unit in the context of neighboring units or areas. The zoning plan specifies the purposes of use or entry allowed or permissible. An Area Statement cannot be inconsistent with the zoning plan, but it can provide more precise specification regarding the level of activity. As an example, reefs may be zoned Marine National Park for the purposes of conservation and of recreation based upon nature appreciation. Without being inconsistent with the zoning plan, at any such reef the level of use could range from a few small boat parties to large daytrip operations and the amenity setting could range from "wilderness" boating and diving to serviced natural history tourism with pontoons and coral viewing vessels. Without more specific guidance there is a risk that all reasonably accessible reefs will gradually approach intensive use and that some forms of reasonable activity will be squeezed out. For intending users, an Area Statement clarifies the intentions of the management agency. Thus at one reef the appropriate setting may be small numbers of small groups seeking a wilderness experience. For another it might be provision for large groups of tourists taking well-managed educational day-trips for an introductory experience of otherwise undisturbed reef. It would be inappropriate to grant a permit for a large tourist operation in the first example, but, subject to impact assessment and available capacity, it would be appropriate in the second.

REEF USE PLANS

A zoning plan provides broad scale control over use and entry to large areas. This may be elaborated by an Area Statement that establishes the desired usage setting. In the case of reefs subject to heavy use, there is often a need for more detailed strategies to address specific management issues. They thus provide a more detailed level of planning to guide users, day-to-day managers, applicants for and assessors of permits concerning site specific issues. Reef Use Plans are problem-oriented and the approach adopted is to state the management issues and then present the management prescription. Reef Use Plans make detailed provision for matters such as the location of moorings, strategies for management of helicopter operations or controlling reef walking, or snorkeling.

Reef Use Plans are prepared and adopted by the Great Barrier Reef Marine Park Authority in consultation with the primary day-to-day management agency, the Queensland National Parks and Wildlife Service. They may be made or revised at any time, but they must always be consistent with the zoning plan provisions for the area to which they apply. The process of plan development involves consultation with key users and groups but does not generally involve full public participation.

Reef Use Plans specify provisions for such matters as the installation of moorings, definition of anchoring areas, the number of tourist operations, the maximum number of people who may visit a site at a particular time, interpretive and monitoring requirements, controls to ensure safe use of boats in crowded areas and arrangements to minimize noise disturbance.

An example is Lady Musgrave Island Reef at the south of the Great Barrier Reef. The island is used for camping and daytrip visits. One major daytrip operator is established there offering an unpressured visit to reef and island. The lagoon is a major anchorage for cruising yachts and fishing vessels. The island is a National Park and an important bird nesting area. Existing levels of use are considered to be sustainable, yet experienced yachtsmen claim that the experience has already been degraded by increased numbers of visitors. The reef and lagoon could, physically, cope with greater numbers of vessels and visitors. The island could not sustain greater visitor levels without encroaching on the bird nesting areas and changing the nature of the island experience. Additional operators would like to take day visitors to the reef and lagoon even if they might not get onto the island. Provided it were possible to prevent visitors from a second major daytrip from going to the island there are no immediate resource-based reasons for refusing a permit although it is clear that to permit a second operation would further change the nature of the experience and create management pressures.

A Reef and Island Use Plan has been adopted for Lady Musgrave which defines the opportunities, conditions, and the setting for use of the reef, lagoon and island within the provisions of the Great Barrier Reef Marine Park Act, the Queensland National Parks and Wildlife Act and the Queensland Marine Parks Act. It makes clear to permit applicants and to permit delegates what will or will not be considered appropriate and should thereby save time, effort, and conflict that might otherwise be involved in design, decision, and appeal over inappropriate proposals.

PLAN REVIEW

Special management areas and Reef Use Plans provide the means to address urgent changes in management issues on a site-specific basis in the short term. In the longer term, as the extent and proportion of uses changes, it becomes necessary to review the zoning plan. The invention or development of new forms of use can change the extent or proportion of activities. Experience from the implementation of management or new knowledge from research and monitoring can give rise to alternative approaches to solving management problems. For these reasons there is a policy that each Zoning Plan is reviewed after it has been in operation for five years.

An example of major change is the development of substantial tourism in the Cairns section of the Great Barrier Reef Marine Park in the five years following the implementation of the first Zoning Plan for the area in 1983.[6] That plan was developed at a time when the demands of the tourist industry were met by vessels traveling at 7–10 knots to a few readily accessible reefs. The first high-speed passenger catamaran, capable of carrying 150 passengers at 25 knots, entered service on the Great Barrier Reef in 1982.

The course of tourist industry development in the Cairns area since the introduction of high-speed catamarans has demonstrated how major changes in use and demand can occur in short periods of time as a consequence of the application of new technologies. The introduction of the high speed catamarans, pontoons, and coral viewing vessels enabled large numbers of day visitors to see the reef. It placed a major new demand on sheltered anchorages with reasonable coral for viewing within 50 nautical miles of Cairns or Port Douglas. It generated opportunities and employment and provided a distinctive attraction as a basis for the development of the transport and accommodation sectors of the Cairns tourist industry.

Permit data returns indicate that day visitor use out of Cairns and Port Douglas rose at approximately 30% per annum between 1985 and 1988. In the same period the number of tourist program permits has increased from 52 to 185 and the number of site-specific permits from 23 to 86. The rapid development of reef tourism gave rise to concern in the community and the industry. For the community there was concern at encroachment upon the amenity of existing recreational and small charter boat usage which gave rise to fears that such usage would be excluded as all possible sites became developed. Sites perceived to be most suitable in terms of access, anchorage and attraction were quickly taken up, and by 1988 it could be foreseen that at some time all suitable sites would be allocated. For the industry there were two avenues of concern. Industry strategists were alarmed at the concept that scarcity of sites might dampen the potential for industry growth. Existing operators were concerned from the perspective of disaster planning. Some saw that if all sites were allocated, there would be little scope for recovery after a natural disaster such as coral destruction by a major cyclone.

In fact, relatively few reefs offered suitable sites with the right combination of accessibility, sheltered mooring site and attractive coral growth. Nevertheless, commercial and recreational fishermen and other small-scale recreational users felt displaced or inhibited from continuing their activities at some of their "best" sites which were now in the vicinity of large tourist operations.

For a few people the concept of large tourist operations and the installation of permanent facilities such as pontoons, observatories or floating hotels is philosophically repellant. Many people were concerned at the possibility that, under the permitting provisions of the zoning plan, all public access reefs could theoretically be "taken over" for large-scale tourism. This introduced the need to avoid real or perceived amenity conflicts by incorporating a tourism strategy into the zoning plan. The review of the Cairns Section Zoning Plan involved introduction of measures to define the areas in which tourist facilities will not be permitted. This provides assurance of continuity of amenity for the range of uses while preventing the emergence of ribbon development with all suitable sites automatically being occupied by major tourist operations to the exclusion of other uses.

In designing zoning plans there is a fundamental choice between a system that makes precise provision for a wide range of activities and one that can be understood by those whom it affects. The initial Great Barrier Reef Marine Park zoning plans were designed around the need to allocate areas in relation to fishing and collecting. There was thus a gradation in restriction in public access zones from General Use Zone A with all forms of fishing including trawling through two or three zones with increasing controls or limitations on fishing and collecting to Marine National Park B Zone in which no fishing or collecting is allowed. Despite the desire to keep the number of zones small, the revised Cairns plan, in making provision for tourist development and no tourist development zones, increased the number of public access zones to eight. Whereas there was appeal in the concept of retaining a simple progression from least to most restricted zones for both fishing and tourism, this was not feasible since there was a need for areas free from fishing and collecting that could be visited by large numbers of tourists for natural history tourism. Equally there was a need for areas in which fishing could take place without the presence of large-scale tourist development. The revision involved creation of a two-dimensional system of controls in which each of the public access zones—the General Use and Marine National Park zones—has two forms. One resembles that of earlier plans in which tourist facilities may be allowed subject to permit, while in the other—the No Facilities sub-zone—permits cannot be granted for such facilities.

In terms of general and imprecisely expressed unease about tourist development on the Great Barrier Reef the adoption of such a strategy should make it clear that the accessible Great Barrier Reef will not be subject to capacity ribbon development and that it will not be used as a source of cheap real estate for tourist activities unconnected with reef activities. The combination of strategy and site planning should define a framework in which tourism can continue to operate as a reasonable use of the Great Barrier Reef Marine Park.

In the future, development of other new uses, such as mariculture or greatly increased numbers of divers leading to demand for specialist cruise boats, are likely to generate new or changed demands. Increased complexity of plan is likely to be the consequence of reviews to accommodate increased complexity of use and potential conflict. The challenge, despite increasing complexity of plans, is to ensure that the contained controls are known to those who are affected. This is addressed by developing maps, leaflets and video materials that detail the controls applying to specific user groups. Those who wish to know all the details may find them from the zoning plan, but for most the controls that apply to groups other than their own are of little immediate relevance, and a relatively simple fishing, diving or boating guide will suffice. Ultimately the success or failure of the Great Barrier Reef Marine Park will continue to depend on user acceptance of the importance and reasonableness of its multiple-use management scheme for marine resources.

Chapter 14

The Galapagos Marine Resources Reserve

The Galapagos Islands are a province of the Republic of Ecuador. A chain of volcanic islands, they straddle the equator some 1,000 km to the west of the mainland of continental South America. Their waters are biologically productive on account of upwellings of nutrient-rich waters carried north by currents from the Antarctic. They and their surrounding waters illustrate many of the issues that face a developing island nation in the management of its maritime environment and providing for economic growth to support a growing human population.

The Galapagos Islands and their waters have a long history of occasional use from pirates in the 16th to 18th centuries, to whalers in the 19th century[1]. The scientific significance of the islands was recognised by Darwin[2] following observations made during a visit to the archipelago in 1835. He was struck by the large number of "aboriginal" or indigenous species. He was "astonished at the amount of creative force... displayed on these small, barren and rocky islands; and still more so at its diverse yet analogous action on points so near each other." These observations of the variation in species separated on closely adjacent islands were important steps in the thinking that led him later to develop the concept of the origin of species.[3]

There is a close nexus between the island ecosystems and the surrounding marine environment. Nesting seabirds and many of the distinctive endemic species of the islands depend entirely upon the marine environment for food. When the productivity of the marine environment is reduced by severe El Nino southern oscillation effects, as in 1982–83, there can be widespread starvation of marine and island animals. Thus, Laurie[4] reported high mortality of marine iguanas.

The global significance of the Galapagos archipelago was officially recognised by UNESCO when it was inscribed as the first entry on the Register of the World Heritage List in 1978. This inscription recognised more than two decades of work by the Ecuadorian Government with the assistance and support of Ecuadorian and international scientists through the Charles Darwin Foundation for the Galapagos Islands (CDF). More than 90% of the land area of the islands was included in the Galapagos National Park in 1959. In the same year the Charles Darwin Research Station was established at Puerto Ayora on the island of Santa Cruz and provided management and research support for the national park while preparations were made for the establishment of the Galapagos National Park Service in 1968.

The issues affecting the Galapagos archipelago and the Great Barrier Reef are very similar. Both cover large areas that contain environments recognized by inclusion on the World Heritage List as being of global significance. Both support fishing and tourism at levels that are economically significant at the local and national level. Both have adjacent human populations that are growing rapidly. Both have been identified as areas that have the potential to attract increased numbers of tourists. For both there is a widely held view that all human activities should be regulated and coordinated to ensure that the nature of their distinctive ecosystems is not significantly damaged by human impact.

THE ECONOMIC SETTING

The population of the Galapagos Islands grew from a few hundred in the 1950s to about 6,000 in the early 1980s.[5] Natural growth is accompanied by immigration from the mainland. The original economic activities of Galapagos Island residents were artisanal fishing, small crop agriculture for local markets and, in the areas outside the Galapagos National Park, agriculture and cattle rearing.

The Galapagos is a substantial contributor of export income to the economy of Ecuador although few of the inhabitants of the islands are directly involved in that activity. The population of the Galapagos is supported to a considerable extent by grants and subsidies from the mainland. There is little potential for increased artisanal fishing. Indeed, performance of the demersal and the lobster fisheries suggest that they may already be exploited at or beyond the maximum sustainable yield.[6] There are industrial scale fisheries for tuna, swordfish and shark but these are operated by high-cost mainland or foreign fishing vessels functionally independent of the local Galapagos economy.

There is technical potential for increased yield and a greater diversity of agricultural and horticultural crops but the cost to transport produce 900 km to the mainland suggests that the role of such activities lies in import substitution rather than in export. The size of the market generated by the resident population is so small that the economic potential of agriculture and horticulture is unlikely to be significant.

The most likely source of sustainable economic support or development for the resident population appears to lie in tourism based on the unique natural attractions of the landscape and the specialized species that have colonized the volcanic islands. Tourism is clearly a factor of great significance for the environmental and economic future of the Galapagos.[7]

Before 1970 tourism was uncontrolled and incidental.[8] It was difficult for foreigners to reach the Galapagos and, without personal contacts on the islands, it was not easy to arrange travel within the archipelago. A number of local residents with boats were prepared to take the occasional party of visitors, but these were not sufficiently predictable to feature even in specialist travel marketing.

By 1970 a number of yachts were operating on scheduled charter for groups of 6–12 passengers; 4,500 visitors were recorded. In 1974, 7,600 tourists visited the Galapagos.[9] The growth of tourism since 1974 is illustrated in Figure 14.1.

Tourism has grown rapidly in the Galapagos archipelago. By the early 1980s, de Groot[10] reported that the system of controls established in the early 1970s had become increasingly stressed as visitor numbers increased. He called for decisive action to limit the growth of tourism. Further substantial growth in visitor numbers occurred after 1986.[11]

MANAGEMENT OF THE GALAPAGOS MARINE ENVIRONMENT

Recognition of the close linkages between the islands and the marine ecosystem led to early proposals that the terrestrial Galapagos National Park should be extended half a mile into the sea to include and protect the intertidal and immediate coastal habitats.[12]

Broadus and Gaines[13] focused on the interaction of marine and coas-tal management and identified problems in the definition and structure of interagency responsibilities and relationships as an obstacle to the development of appropriate procedural frameworks for policy and management. They discussed a scheme proposed earlier[14] in which the waters surrounding the islands are a single marine special management zone and the surrounding territorial sea is an additional external marine buffer. They noted that the marine special management zone could be organized to permit multiple use activities and to confer protection to parts of the marine environment.,This scheme foreshadowed at least three management subunits—Buffer, Special Management (General) and Special Management (Protection).

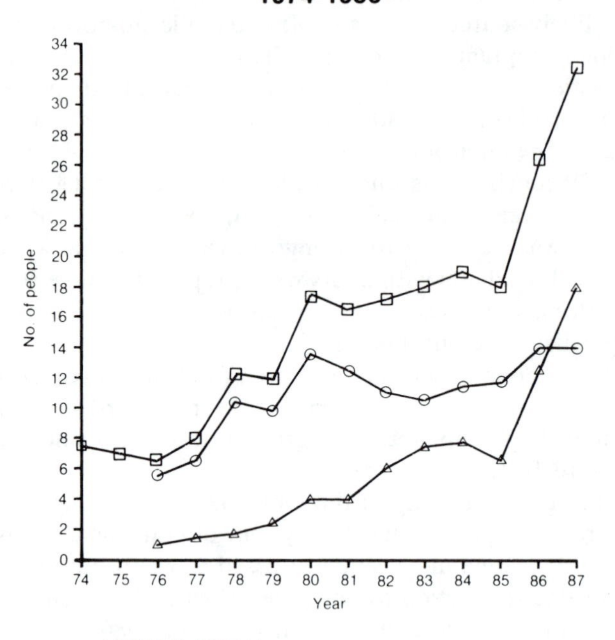

Figure 14.1 Graph showing visitor numbers to the Galapagos Islands in thousands from 1974 to 1987.

In April 1986 while the work of Broadus and Gaines[15] was being published, the Galapagos Marine Resources Reserve (GMRR) was proclaimed in Presidential Decree 1810-A. This established a marine resources reserve covering the seabed and waters within 15 nautical miles of the internal baseline around the islands of the Galapagos archipelago. The decree established an Inter-Institutional Commission of Government Ministers or their representatives to work with national and international agencies to develop arrangements for administration, management, development, and control of the Galapagos Marine Resources Reserve.

By late 1987 a draft zoning plan had been prepared with an accompanying report suggesting approaches to resolving a range of legal, policy, resource, training and logistic issues in order to establish effective and integrated management of the Galapagos Marine Resources Reserve and Galapagos National Parks Service. This stage was reached by a process that involved a high degree of interdisciplinary, national and international collaboration and support.[16]

THE PROCESS OF PLAN DEVELOPMENT

In late 1986 the Inter-Institutional Commission appointed a fourperson Technical Commission (TC) of experts seconded part-time from four government agencies:

- Instituto Oceanografico de la Armada (INOCAR) Naval Oceanographic Institute
- Servicio Parque Nacional Galapagos (SPNG) Galapagos National Parks Service
- Instituto Nacional de Pesca: National Fisheries Institute
- Instituto Nacional Galapagos (INGALA) National Institute for the Galapagos

At the same time four international groups were approached to provide expert advice and support through membership of an International Commission:

- The Great Barrier Reef Marine Park Authority, Australia (GBRMPA)
- The National Oceanic and Atmospheric Administration, USA (NOAA)
- The University of Rhode Island, USA (URI)
- The Woods Hole Oceanographic Institution, USA (WHOI)

The International Commission never met, but individual institutions undertook specific tasks. WHOI undertook the role of coordinating a review of relevant physical, biological and socioeconomic data. It produced a special issue of the magazine OCEANUS[17] devoted to the marine environment of the Galapagos and specifically to the Galapagos Marine Resources Reserve. In April 1987 with funding obtained from the National Science Foundation of the US, the Marine Policy Center of WHOI organized a major workshop in Guayaquil. Expert reviews were presented on issues relevant to the development of a management plan for the Galapagos Marine Resources Reserve. The workshop adopted a series of recommendations. The proceedings of this workshop are being published in English and Spanish.[18]

NOAA, as part of its International Training and Advice Program, seconded an officer to work with the Technical Commission for three weeks in July and August 1987 on the definition of management issues, opportunities, and constraints for the Galapagos Marine Resources Reserve.

GBRMPA, using funding provided through the Heritage Program of the Ecology Division of UNESCO, seconded an officer to work with the Technical Commission. The task of this phase of the planning process was to draft a management plan, propose regulations, and provide a report to the IIC. A subsequent secondment from WHOI provided assistance for the TC in the task of revision of the proposed plan, regulations, and report to take account of comments raised by the IIC in consideration of the initial report and plan.

ISSUES OF PLAN DEVELOPMENT

The problem analysis conducted by the Technical Commission with advice from Lemay explored the interaction of uses and impacts. It identified a number of general existing and potential usage and resource conflicts. It also identified a number of specific problem sites where management decisions would be required to resolve differing and incompatible requirements among fishing, tourism/recreation, and science/conservation.

Wellington[19] had proposed two types of reference sites free from the human impacts of harvesting natural resources and managed consistently with the terrestrial national park. In addition to providing some protected areas for exploited species these sites could serve one of two purposes. One would be as sites available for nonextractive recreation and tourism but free from the immediate effects of fishing or collecting. The other would be as refuge sites free from human access, corresponding to refuge and scientific areas of a terrestrial national park. In Wellington's proposal these sites correspond to intensive and extensive use sites in the terrestrial national park.

Clearly, an effective plan would need some form of area subdivision and activity allocation to provide for a range of competing and sometimes incompatible activities. In developing a draft plan, draft regulations, and accompanying report, the Technical Commission drew heavily on the model of zoning developed by the GBRMPA.[20]

The first step toward development of the zoning plan was to build a logical framework to guide decision making. The goal and aims of the Great Barrier Reef Marine Park Authority[21] were adapted for the Galapagos Marine Resources Reserve. The guidelines for development of zoning plans[22] were similarly adapted.

The second step was to discuss the management needs and the planning process with Galapagos residents and others who would be directly affected by introduction of management of the Galapagos Marine Resources Reserve. During August and September 1987 meetings were held with fishermen and local government representatives at the main population center, Puerto Baquerizo Moreno, on San Cristobal Island and with local residents of Puerto Ayora, on Santa Cruz Island. The Technical Commission continued this series of meetings in October and November so that all settlements on the Galapagos Islands were visited. Meetings were also held in Guayaquil and Quito to brief officials of government departments and to seek their comments on management issues and priorities.

An information summary was developed, largely on the basis of papers presented at the workshop organized by the WHOI in Guayaquil in April 1987[23] and the outcome of discussions with users and government officials. The information was summarized in the form of a series of maps, figures and tables. These provided qualitative and quantitative information on the distribution and abundance of resources and of human uses and impacts. The goal, aims, and guidelines for plan development were then applied, using the best available information to develop a draft zoning plan. The draft plan, regulations, and report arising from this process were presented to the Inter-Institutional Commission for initial discussion in September 1987.

In November 1987 the plan, regulations, and report were revised by the Technical Commission to take account of comments, criticisms and specific technical issues raised by the Inter-Institutional Commission and by WHOI and NOAA specialists.

THE PROPOSED ZONING PLAN AND REGULATIONS

Each part of the area of the GMRR is allocated to one of four zones. The zoning plan specifies the uses that may take place in each zone. Any activity not specified may not take place, although if it is consistent with the objectives of the zone, it may be allowed by permit.

The objectives of the zones are:

1. Zona de Uso General (General Use Zone): To provide opportunities for sustainable general use consistent with the conservation of the Galapagos Marine Resources Reserve
2. Zona de Pesca Artesenal y de Recreacion (Artisanal and Recre-

ational Fishing Zone): To provide for the protection of productive areas of the Galapagos Marine Resources Reserve while assuring opportunities for the continuation and growth of artisanal fishing for the benefit of residents of the Galapagos archipelago and opportunities for appreciation and enjoyment, including recreational fishing.
3. Zona de Parque Nacional Marino (Marine National Park Zone): To provide for the protection and conservation of areas of the Galapagos Marine Resources Reserve, while allowing opportunities for their appreciation and enjoyment by the public free from activities that remove natural resources
4. Zona de Reserva Intangible (Strict Nature Reserve): To promote the preservation of critical areas of the Galapagos Marine Resources Reserve in their natural state undisturbed by human activities.

A map illustrating the proposed zoning plan is shown in Figure 14.2.

MANAGEMENT OF THE GALAPAGOS MARINE RESOURCES RESERVE

The creation of a zoning plan is the first and easiest stage in management. The approach of multiple-use management that has been adopted depends upon coordination of management whereby existing agencies retain their established management responsibilities. Such a system requires that provision be made for effective consultation and coordination in order to establish agreed policy and priorities for management and, where necessary, sharing of resources in order to carry out agreed programs of all government agencies with responsibilities within and adjacent to the area of the Galapagos Marine Resources Reserve.

At the time the revised zoning plan was presented to the IIC the coordination mechanism was not established. The IIC determined that SPNG should be the primary managing agency for the National Park and Strict Nature Reserve Zones and INGALA for other zones. However, because of rundown of staff numbers and the linked lack of functional patrol boat capacity, none of the agencies had, at that time, the capacity to meet its existing responsibilities beyond port limits, let alone to undertake new responsibilities. In the climate of extremely scarce resources, it appeared unlikely that any of the agencies could individually acquire the necessary staff and financial support to discharge its responsibilities independently.

A coordinated approach appeared to offer the only financially feasible possibility of acquiring the capacity to meet existing agency objectives and the new requirements arising from the Galapagos Marine Resources

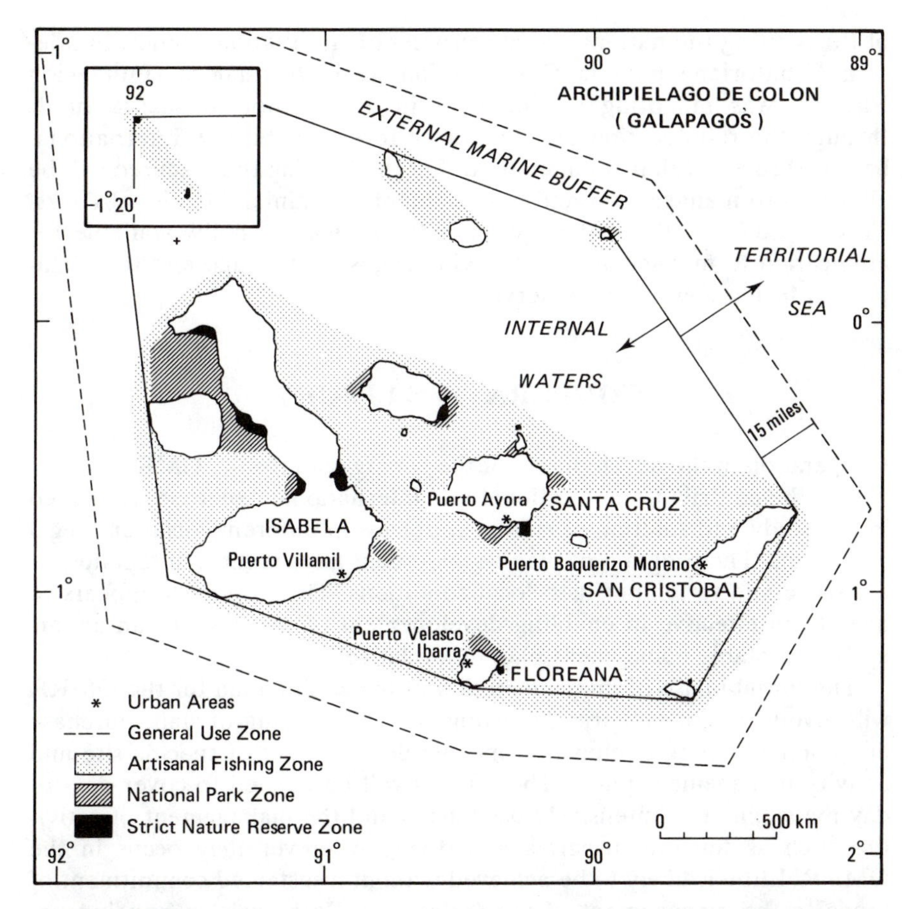

Figure 14.2 Map showing the proposed zoning of the Galapagos Marine Reources Reserve.

Reserve. However, the benefits of the ability to meet existing and new responsibilities will come at the cost of some loss of apparent autonomy to each of the collaborating agencies. It remains to be seen whether it will be possible to establish such a coordinated approach.

The critical issue is that of establishing a management system with a secure, self-sustaining funding base. Visitor fees levied from international visitors upon arrival at airports in the Galapagos Islands yield sufficient funds for the purpose. However, at present, in the absence of equivalent funding sources for the rest of the Ecuadorian National Parks system, the majority of the Galapagos levy is apparently applied to meet urgent basic requirements of mainland national parks management. At the time of this writing the national nongovernment conservation agency, Fundacion Ecuatoriana para la Conservacion de la Naturaleza (Fundacion Natura), was mounting an initiative in an attempt to secure funds through the re-allocation of Ecuadorian foreign debt with international banks. If successful, it is understood that the funding thus secured will be allocated to management and development of mainland National Parks. This would leave the Galapagos levy theoretically wholly available for application to management of the Galapagos archipelago and the Galapagos Marine Resources Reserve.

COOPERATIVE EFFORT

Cooperation in the development of the zoning plan for the GMRR has involved the contribution of national and international expertise developed in the study and solution of similar problems in different social and legal conditions. The program of plan development has sought to integrate this expertise and to temper it to local conditions. This approach appears to have been effective in enabling developement of the basis of an apparently acceptable managment plan in a relatively short time.

The initial phase of implementation of the zoning plan for the GMRR will involve major activity; the acquistion and training of staff, purchase and commissioning equipment, and the development of specific site and activity management plans. These plans will be needed to cover day-to-day managment of intensively used areas and the management of activities such as tourism, fisheries and diving wherever they occur in the GMRR. Little is likely to be achieved without a sustained commitment of funds to the management of the Galapagos. With such an ongoing national committment to long-term management, it is to be hoped that international support and expertise can be applied to provide training, equipment, research and monitoring, and thus establish a secure basis for long-term managment of the Galapagos archipelago.

CONCLUSION

Planning for the managment of the Galapagos Marine Resources Reserve has proceeded through the participation of a remarkable range of national and international agencies. The contribution of international and national agencies have reflected the continuing acceptance of the importance of the Galapagos archipelago in the context of the natural heritage of the world. It is clear that the urgent need for coordination of activities identified for management of the marine environments of the Galapagos applies equally to their terrestrial closely linked island enviroments. It is to be expected that the current pressures in the Galapagos will increase. It is a matter of urgency that an effective system of managment coordination be established and securely funded for effective implementation of measures developed to provide for long—term conservation and reasonable use of these unique environments.

Chapter 15
The Republic of Maldives

The Republic of Maldives is a developing oceanic island nation with no mainland continental territory. Situated in the middle of the Indian Ocean it consists of a chain of 23 atolls no more than 130 km wide and 800 km long, extending from just south of the equator to just north of 7°N. The atolls contain complexes of sandbanks, bars, and coral reefs. Depending on the definition of a reef there are approximately 5,000 coral reefs, about 1800 of which carry small vegetated coral cays or islands. About 200 of the islands are permanently inhabited. Many others are visited for short periods for harvesting coconuts, occasional cropping and collection of firewood. The Republic is administered through 19 administrative units, most consisting of a single atoll or, part of a large atoll. These are indicated by the letters A to S on Figure 15.1.

The total population in 1985 was approximately 180,000 people, 46,000 living on one island in the capital city of Male. The population is growing rapidly. From the beginning of the 20th century until 1960, it was stable fluctuating between 70,000 and 80,000. It was 130,000 in 1974 and is projected to reach 280,000 by the end of the century.[1] All the material of the reefs and islands is of biological origin. The natural resource base to support the existing and growing human population depends upon the well-being of the marine environment.

ECONOMIC SETTING

All the natural resources of the Maldives are derived from the sea. Until quite recently, the skipjack tuna was virtually the sole basis of the economy. It provides the main dietary protein of the Maldivian people. Historically exported as Maldive fish, a traditional food for those on the *Haj*, or Islamic pilgrimage to Mecca, smoked and dried skipjack tuna provided in-

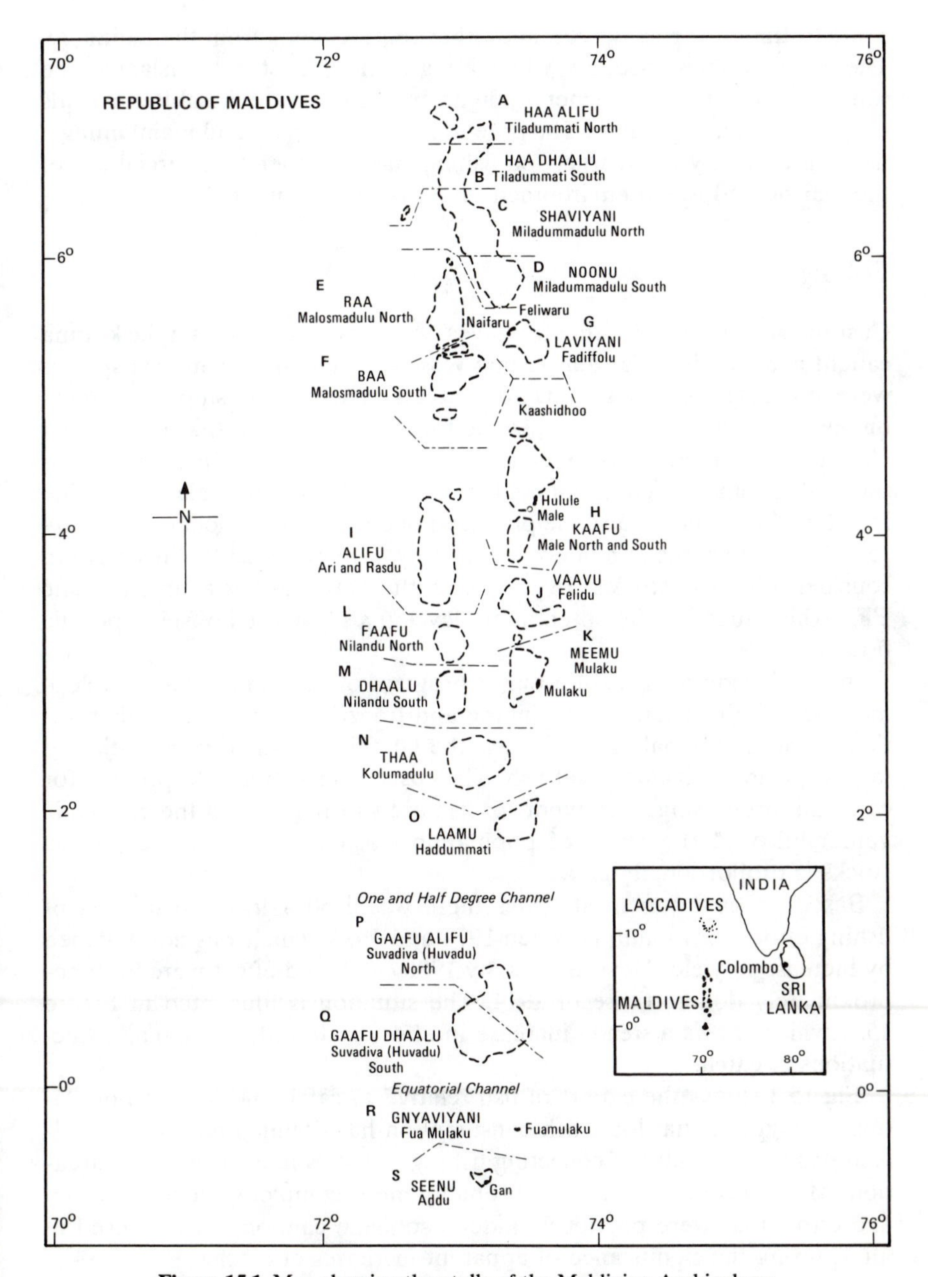

Figure 15.1 Map showing the atolls of the Maldivian Archipelago

come to trade for rice, spices, and other requirements from the mainland. The other staple is coconuts, which are grown on most of the islands. The soil of the coral cays is generally light, but most of the inhabited islands support vegetable gardens. The potential for developing and maintaining a modern economy rests largely on fishing and the other commercial use of the marine and island environment, international tourism.[2]

Fishing

Customarily, fishing in the Maldives was based on the skipjack tuna caught near the islands from sailing vessels, or *Dhonis*. Demersal species were not regarded as serious food items. The relationship of a small human population to a large pelagic fish stock was such that fish was an abundant commodity to meet local needs and through export sale or trade, dried and salted or smoked, to provide for other basic needs. The increase in population and the pressures of developing a modern economy to share the benefits of modern technology have provided the incentive to consider other fish stocks.[3] In particular, the Maldives has a large oceanic EFZ which suggests the potential to develop a substantial offshore pelagic fishery.

In the decade from 1974 a major program of investment and development saw the installation of canning and freezing works, the mechanization of the traditional fishing vessels the Dhonis and the diversification of target species to include reef fish. The costs of investment to provide for catching, processing, and export of fish are substantial and the economic vulnerability in the event of population crash of any of the exploited stocks is proportionally great.

Based on government statistics there was a 50% increase in days of fishing effort per annum between 1974 and 1983, which was not matched by increase in yield. Nevertheless, by 1986 catch and effort were both approximately double those of 1974. The situation is illustrated in Figure 15.2, which shows a steady increase in effort with underlying major fluctuations in catch.

Fig 15.3 shows the export of fish relative to catch and human population. It suggests that local fish consumption has doubled but probably illustrates the difficulty of collecting fishing statistics in a subsistence situation. More recent statistics are probably more complete, including catch and effort that were previously hidden so some caution must be used in interpreting the significance of apparent increases of catch.

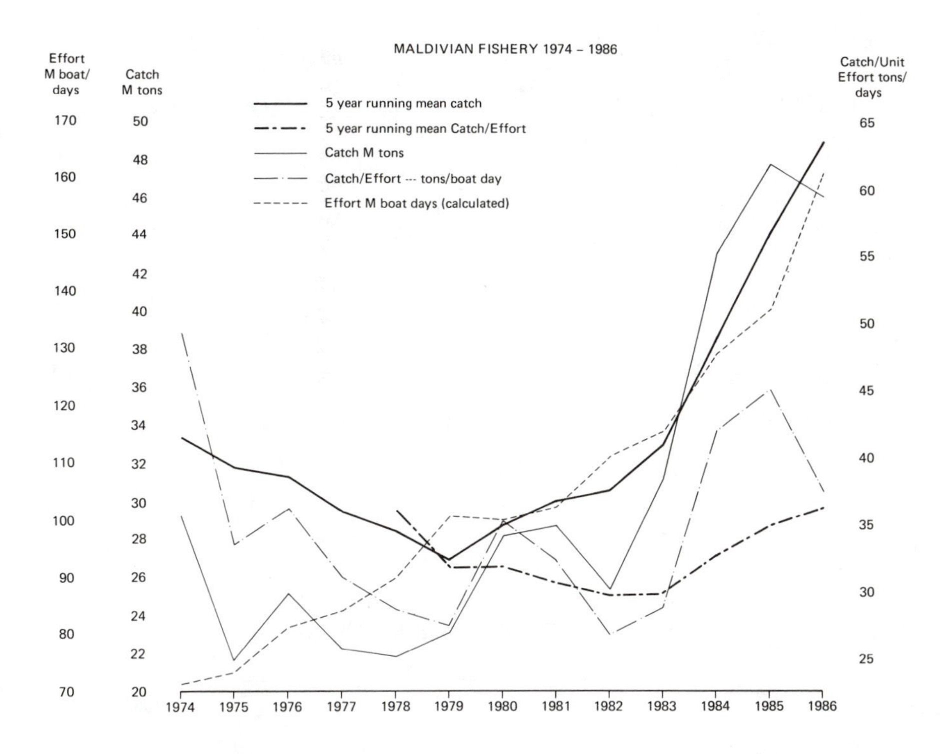

Figure 15.2 Graph illustrating the Maldivian fishery between 1974 and 1986. Compiled from the Maldivian Statistical Records.

Tourism

The rapid development of the tourist industry was the result of careful planning to take advantage of an ideal economic opportunity for a developing nation. Compared with many other economically developing nations, the Maldives has a particular advantage as an area for tourist development. The large number of uninhabited islands provide opportunities for the creation and operation of tourist resorts without significant disruption of social patterns or displacement of existing inhabitants.

The main problem at the start, which affected most economic development possibilities, was that of air access. Few of the islands are large enough to accommodate an airstrip for light aircraft let alone to meet the requirements for international jets. Gan Island, in Addu Atoll in the extreme south of the country, has an airstrip which was created by the

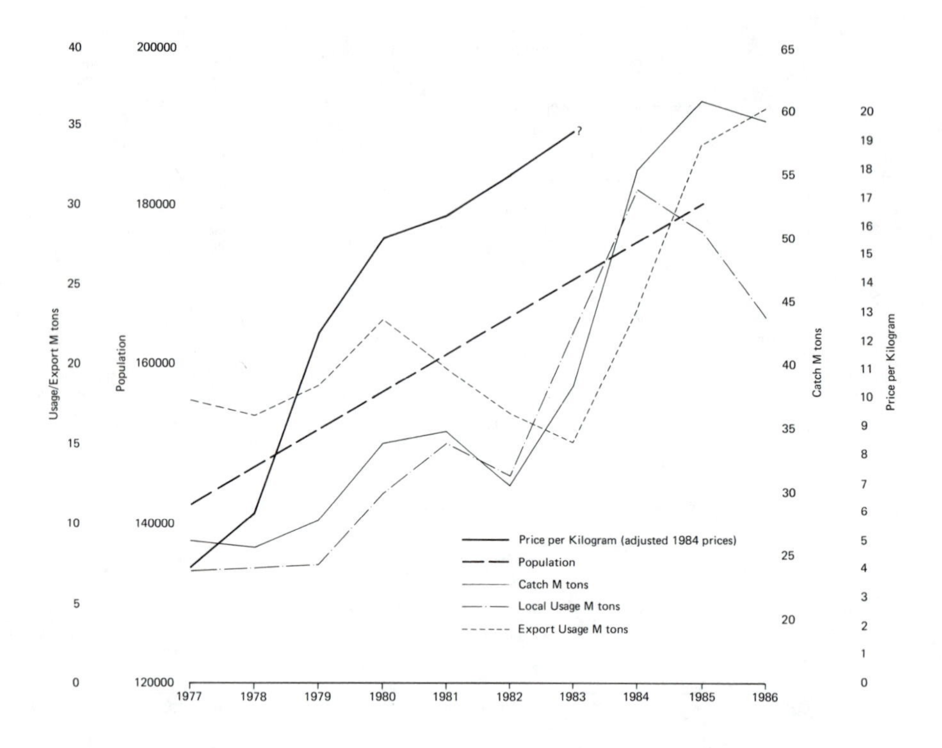

Figure 15.3 Graph showing the relationship between catch, usage and price of fish and growth of human population, 1974-1986.Compiled from Maldivian Statistical Records.

British as a military communications base, decommissioned in 1970. That airstrip was not large enough to meet commercial operational requirements for Boeing 707 jets. Although it could be extended to meet those requirements, it would have been very difficult and expensive to provide further extension to receive widebodied jets. In any case Gan is almost 300 miles from the capital city, Male, and 500 miles from the northernmost atoll of the republic so substantial investment to create the main international airport there would be difficult to justify. The critical decision was to commit substantial resources and to seek international assistance to create a modern airstrip close to the capital. This was achieved by dumping large amounts of fill to link Hulule and an adjacent small island. Widebodied jets were first able to land at Hulule in 1981.

The Hulule airport opened the way for some 50 resorts to be developed

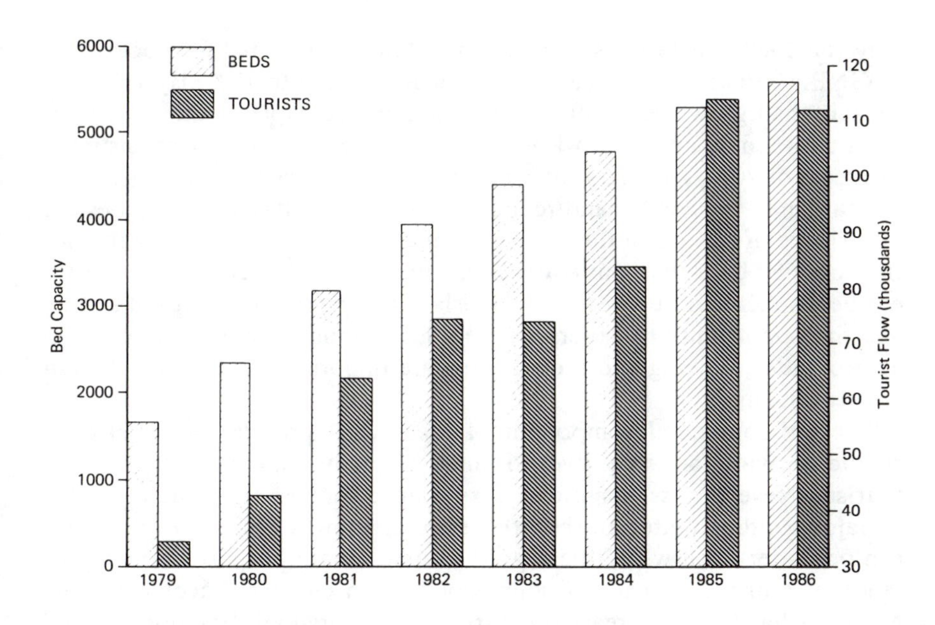

Figure 15.4 Histogram showing relationship between available tourist beds and visitor arrivals to the Maldives, 1979–1986. Compiled from Maldivian Statistical Records.

within reasonable boat access. Development was coordinated by the government largely through the Department of Tourism. There was an early decision to limit foreign ownership, although foreign involvement in management was seen as a way of developing standards and expertise. The capacity for further development is related to the introduction and maintenance of large capacity rapid passenger vessels or of the development of other airports to receive feeder services from Hulule.

Government policy has tightly controlled the location of tourist developments. Most atolls are off limits to foreign visitors without specific permits and, as such, not available for tourist development. When an atoll is available for tourist development only selected islands may be developed and the standard of development is controlled by the Department of Tourism.[4] Initially that control related mainly to quality of finish and presentation, but as a result of unsatisfactory experience it has extended to environmental issues of waste and sewage disposal and to erosion control.

Tourism statistics are reported from 1972 when there were 280 resort beds in operation. By 1980 there were 2,418 beds, and the foreign exchange earned exceeded that of fishing. In terms of gross domestic product between 1980 and 1986 fishing has consistently accounted for about 20% of GNP whereas tourism has grown steadily from 9 to 18%. By 1986 bed capacity had grown to 5,559. Average annual bed capacity utilization has been between 40 and 55% with rates for the January quarter consistently in excess of 70%, February rates are usually in excess of 80%.

Tourism to the coral island/reef complexes of the Maldives provides the experience of aspects of the natural environment. The major marketed attraction may be as basic as warm sea, sun, and sand in contrast to northern hemisphere winter weather. The islands also provide a natural setting and the opportunity to experience and learn about complex and colorful environments. Diving and reef viewing are important parts of the Maldivian tourist market.

The environmental component of the tourist product or experience should provide an incentive to limit the environmental damage of tourism. Nevertheless, some early experience showed that where coral is a major building material, the process of creating a tourist development can be undertaken with little regard to its ultimate purpose. The importance of tourism as a major source of foreign currency income for the Maldives has led to increasing control and the gradual development and refinement of standards for tourist development and operations.

Manufacturing

The only significant sector of the export economy that does not rely on the natural resources of the Maldivian archipelago is the manufacture of clothing, the other major export economic activity. Established in 1983, it accounts for about 20% of export income, although textile imports offset about half of these. Beyond this, there is scope for additional manufacturing industry which is likely to be explored as population growth leads to increased demand for employment and economic opportunities.

MANAGEMENT OF THE MARINE ENVIRONMENT

The Maldives is a nation with a history of Islamic religion and government dating back to 1153. Whereas the highest levels of government structure have changed from a sultanate to a written constitution with a presidential administration and a parliament, or *Majlis*, the structural hierarchy of local government has changed little. The capital island city of Male is a special administrative unit that combines the general problems that capital cities of developing countries face in providing infrastructure for high-density living with those of creating and maintaining the land area to support that infrastructure. The other 18 administrative "atolls" retain more of the traditional hierarchical approach to managing relatively small human populations on islands scattered through a large area. Most comprise single physical atolls, although some of the larger one have been subdivided to two administrative atolls and some smaller physical atolls have been amalgamated for administrative purposes.

Each administrative atoll has an atoll chief, a senior local government political/administrative figure. The atolls are subdivided into administrative units which usually comprise a populated island and surrounding supporting area of fishing grounds and uninhabited or plantation islands. Each administrative unit has an island chief. The atoll chief and island chiefs form a council which handles day-to-day administration throughout the atoll. The chiefs are supported by a form of public information and consultation which takes place through formal or informal meetings often arranged by the *Mullah* to follow worship at the local mosque.

Decisions on economic and environmental matters such as use or closure of sites for fishing or collecting of shells or corals are apparently governed by long history with specific issues being decided through the island/atoll chief council and decisions once made being implemented by announcement and community acceptance. New measures are becoming necessary to manage environmental issues that arise as growth of the human population and the level of economic activity have increasing potential to move beyond the sustainable capacity of the environment. Provided the management decisions can be made and generally agreed through the established process the system would appear to lend itself to the development and implementation of such new measures.

THE PROCESS OF PLAN DEVELOPMENT

In 1984 the government of the Republic of Maldives sought assistance from UNESCO because of concern over the condition of the environment and the need for management. The government identified nine areas of concern regarding the environment and the need for its management:

- Pollution
- Littering
- Tourism
- Human environment and health
- Effects of reclamation
- Coral mining
- Causeway construction
- Fishing
- Bait fishing

In 1984 UNESCO arranged for a short consultant study to review the conservation status of the coral islands and marine environment of the Maldives and to make recommendations regarding a research and training program for environmental management. The study was required to take into account pressures from tourism, fishing, pollution and mechanical destruction, and current conservation practices and law. The study reported on the concerns identified by the Government and developed initial proposals for multiple-use management through zoning. A follow-up study in 1985 involved discussion of proposals with government officials, a specific review of erosion problems in Addu Atoll in the south of the Republic, and the refinement of a program of research needs.

ISSUES OF PLAN DEVELOPMENT

Pollution

Clear signs of pollution were observed in the boat harbor, around the outer breakwater and at other locations around Male Island. These were associated with a range of other disturbances arising from massive reclamation and the physical problems of providing for a population of over 30,000 people on a small coral sand island.

Elsewhere in the two atolls visited there were symptoms of pollution in some locations. These took the form of high algal cover, patchy coral death, and high populations of detritus-feeding animals. The conclusion was that the symptoms at areas outside Male Island were not generally so intense or widespread as to cause serious concern but that the situation should be monitored.

The need to identify and exclude toxic materials from reclamation dumps was identified as particularly important since such dissolved substances could easily enter the freshwater lens beneath the islands.

Littering

Littering was observed in several locations on reefs near populated islands and tourist r,esorts. Disposal of non or slowly degradable packaging materials—cans, bottles and plastics was the main problem. It was not generally at a level to cause serious concern for ecological impacts but it was seen as a problem that degraded environmental amenity. It could have significant economic impacts, particularly in tourist areas, if it is not managed. If the value of clean solid wastes as materials for construction fill is widely understood, this may in small part address the problem.

Tourism

Plan development issues relate to structure, process, and amenity. The structural issues of site selection are largely addressed in the consultative process administered through the Department of Tourism. There may be further structural conservation issues where significant amounts of coral and coral rock are used in construction. This has been a particular problem in areas of Male Atoll, but by 1984 it was recognized as a problem to be addressed in the process of development approval by the Department of Tourism.

There are two sets of process issues. The first relates to standards of design and operation of facilities for waste and sewage disposal. The second relates to environmental engineering design of structures such as jetties, breakwaters, and groynes. There are many examples of such structures blocking natural sediment, tidal, or current flows. Where they alter the processes of sediment movement responsible for growth or maintenance of islands and beaches, they may cause erosion in some areas or unwanted accumulation of sand in others. Where they block or alter current flows, they may disrupt migratory paths of fish and other animals.

The amenity issues arise from the displacement of customary uses of islands and reefal areas. The process of planning coordinated through the Department of Tourism addresses the island issues. The reef issues are generally not substantial. Reef use for diving and other water sports has little impact upon customary fishing for pelagic fish species. If fishery development turns strongly to taking reef demersal or territorial species, an amenity issue could arise if stocks of large and attractive fish at dive sites are seen as available fish resources for a fishery.

Human Environment and Health

The critical issues are those of a clean fresh water supply and proper disposal of sewage and waste. In the smaller island communities rainwater collection and limited ground water collection from wells provide reasonable water and the volume of wastes and sewage is generally within the natural absorption and processing capacity of natural oxidation and bacterial activity. In the larger communities and particularly the capital of Male, these problems interact and their solution depends on complex and costly engineering.

When water supply from rainwater collection is inadequate the first step is usually to augment it by pumping from the freshwater lens that exists under most stable coral cays. In essence the lens forms through a chemical interaction within the calcium carbonate sand at the point where rainwater that has percolated downward meets seawater. The result is a form of natural cementation that binds sand, gravel, and other materials into a brittle aggregate. When exposed by movement of the cay, the aggregate can be seen as beachrock. Beneath a cay the aggregate can form an imperfect, but generally effective barrier at the interface between fresh and salt water and so provide a store of freshwater. It is this process that can provide the basis for development of substantial vegetation including large trees. The extent of the freshwater lens depends on a number of factors that relate to the stability of the cay and the amount of rainwater that percolates through the sand to maintain the process and replenish the freshwater supply.[5]

The development of a large human population on a coral cay results in a demand for water that restricts the flow into the lens, and if it also results in substantial drawing upon the stored water, it can further affect the process as the interface between salt and freshwater rises. As the human population rises, the disposal of wastes and particularly of sewage is likely to exceed natural processing capacity and this may result in harmful bacteria entering the groundwater. If this is accompanied by the development of large buildings that require substantial foundations, the process may be further distorted if the brittle rock of the lens interface is fractured. The problem can be addressed by installation of sewage systems but the instability of coral cays makes it essential that pipework be sufficiently flexible to cope with movement of substrate.

Erosion, Cay Dynamics, and the Effects of Reclamation

Coral cays are dynamic structures; they grow by accumulation of coral skeletons and other biogenic remains deposited on the top of a reef by the action of waves. They are moulded by the action of tide and wind-driven current, and they may move or be removed by strong wave action in severe storms. They are stabilized by vegetation, but generally substantial vegetation can become established only on an area of a cay that is rarely affected by wave action.[6]

There is a strong linkage between the human health issues discussed above and the effects of reclamation. Reclamation occurs as the human population creates more land, alienating intertidal or subtidal areas by dumping spoil to raise those areas permanently above the high water mark. The immediate effect is to alter the pattern of wave movement around the island or reef on which the alienation occurs. Generally the new land is in an area of higher wave energy than pre-existing land, and therefore measures are required to prevent its erosion. These will usually take the form of a breakwater or sea wall to reflect or refract the waves. Without careful design this may merely move the problem to another part of the island.

A second effect may arise from the changed or increased load of materials on top of the reef structure that support the island. Reefs are inconsistent structures of limestone skeletal material from corals and other biota ranging from massive coral heads to fine silt. On the surface they are cemented by corals and algae, but attacked by boring animals and algae. Deeper, they may be bound by metamorphosis and cementation of calcium carbonate but the structure is a mixture of substantial framework with spaces filled by sand or gravel-size fragments. Substantial reclamation, particularly when it is accompanied by construction of multistory buildings, may alter the distribution of the mass of materials on top of the supporting reef, and this may lead to movement or slumping of the old or new areas resulting in cracking of foundations and fracturing of water or sewage pipelines.

A third effect may occur if the fill material contains toxic wastes. Clearly there is a risk that such wastes will make their way into the forshore environment or be carried in solution into the freshwater lens. Prevention of such problems requires careful control of types of material to be accepted as fill.

Coral Mining

Coral reefs produce the only naturally occurring rock material in the Maldives. Coral rock, the framework of the current reefs has traditionally been used as fill. Solid coral heads, split with wedges, form the basic natural building block, whereas branching coral burned in a hot fire provides the lime need to make cement.

Until quite recently most buildings were made of timber, and coral was reserved for prestige public buildings, mosques and the residences of high officials. This limitation ceased to apply from the middle of the 20th century and coral has been widely used for much of the building that has accompanied the economic development of the Maldives.

The problems that arise from coral mining are that the living coral and the underlying coral rock substrate are removed. Large areas of reefs are devastated by the process with the consequent loss of habitat for species that depend on corals. The problem is worse close to the developed islands and worst of all within Male Atoll. It extends to many of the other atolls since building materials may be imported and also because some goods-carrying vessels visiting Male are required to pay harbour dues in the form of quantities of coral fill or building materials.

The issue of coral mining may be the most serious immediate environmental problem. A solution will be costly and will involve a major program of public education in order to introduce and achieve acceptance of alternative materials.

Concrete blocks can be made using coral sand but if the sand is not well washed to remove salt, the blocks are likely to crumble. Blocks are increasingly used but they are regarded as inferior materials. In high quality buildings they are usually covered by a cladding material that may be small coral blocks. Technical research, which led to a means of making reliable robust building materials using unconsolidated sand from the deeper waters of the lagoons, would solve several environmental and economic problems.

Causeway Construction

In a nation consisting of small islands, linking adjacent islands by causeways can provide improved communication and access to public facilities and opportunities for economies of scale, for example, by having one larger school or health center rather than multiple small ones. The problems are similar to those that may be caused by jetties, breakwaters, and groynes but potentially on a greater scale. Unless they are properly designed and constructed, causeways can cause substantial, rapid and irreversible movements of beach sediments, resulting in erosion in some areas and accumulation in others. An example is illustrated in Figure 15.4, which shows apparent erosion and accretion areas resulting from the linking of the western island of Addu Atoll by a causeway.

Proper design depends on a good understanding of the pattern of water movements in the range of tidal and wind conditions that can apply to the area. it also involves creating structures that will achieve their intended purpose with the minimum possible disruption of normal water flow.

Fishing

The environmental management challenge of fishing in the Maldives is the same as that of any other nation for which fishing is a substantial source of food, employment, and export income.[7] There is pressure to maximize development, and economic returns by increasing fishing without a clear understanding of the nature and extent of the resource. At the local level on islands where pelagic fish had been the staple, readily available food for generations, the concept of recruitment overfishing and population collapse is not readily understood since it appears to be an alarmist fantasy. The management challenge is to develop fishing efforts within the bounds of the sustainable capacity of the stocks and to avoid overcapitalization, which could in times of lower yield lead to financial pressure to maintain export levels at the cost of local subsistence. Failure to meet this challenge

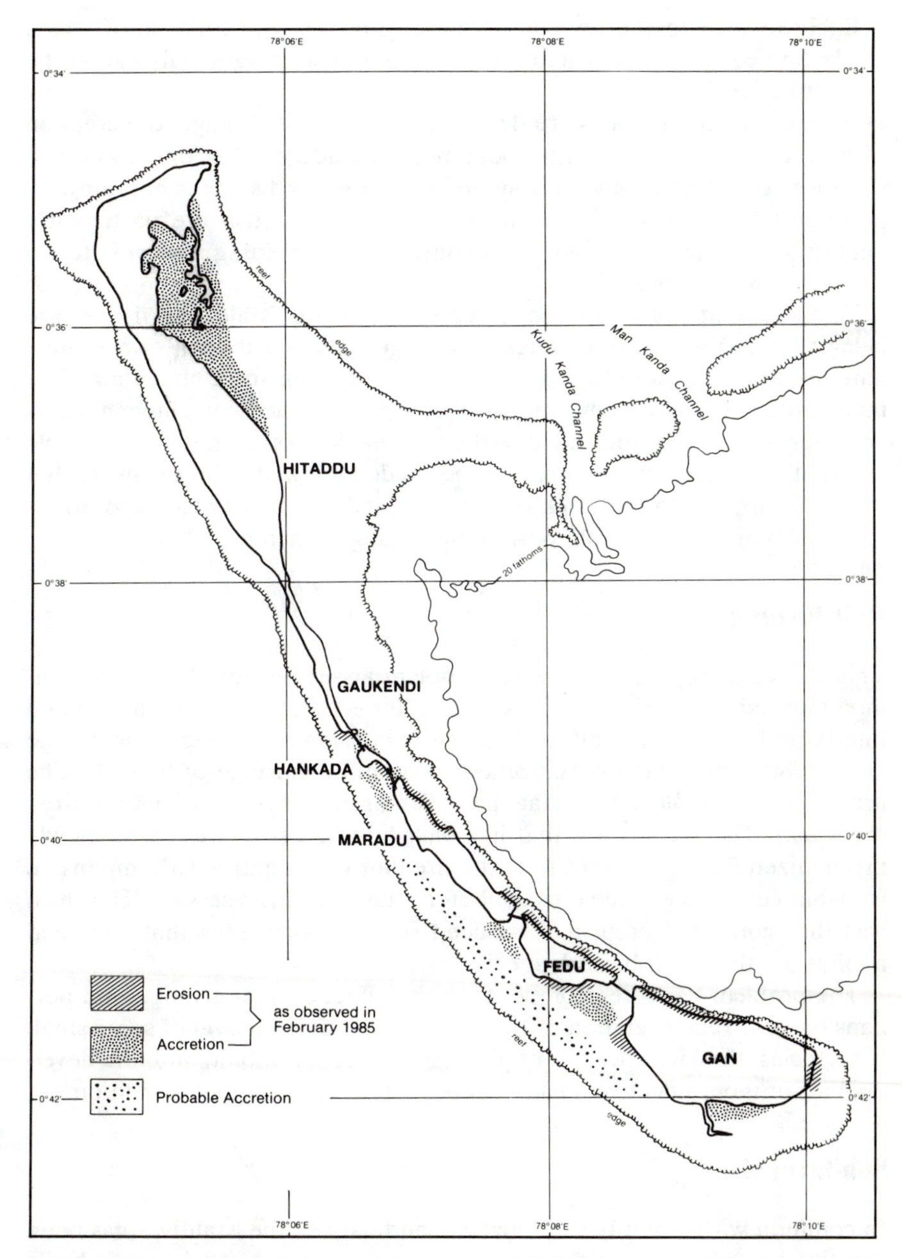

Figure 15.5 Map showing erosion and accretion of sediments apparent following the construction of solid causeways at Addu Atoll Republic of Maldives.

is likely to lead to the tragedy of the commons[8]—the collapse of fished stocks and consequent substantial economic hardship nationally and at the household level.

Where the target stocks are demersal with a limited range, the creation of reserves or national park-type zones may address the uncertainty of fisheries. This may involve setting up areas in which fishing is not allowed, permanently or during breeding periods when the fish are particularly vulnerable. It may contribute by maintaining a breeding reservoir to re-stock the fished population.

To the extent that the main fished stocks of the Maldivian fishery are pelagic species with a large migratory range, the contribution of a zoning, national park, or Biosphere reserve scheme is even more uncertain. The major role may be the educational one of helping the economic managers and users of the fish stock address the fact that stocks are not inexhaustible and that management measures will be needed at some point to ensure that the effort applied to the stock does not exceed its capacity to breed, to restore its numbers and to support an increasing human population.

Bait Fishing

This subset of the wider fisheries problem arises because the most common method of catching skipjack tuna in the Maldives is to use a pole and line baited with live small fish. In the expansion of fishing effort in the early 1980s, three factors combined to create a shortage of baitfish. The first was the simple effect that more fishing in longer sessions required more bait. The second was that live bait did not last so well in the newly mechanized fishing *Dhonis*, so still more bait was required to compensate for what could be expected to die before the end of the session. The third was that coral collecting had reduced some of the areas that provided habitat for the small baitfish species.

The problem has three components: the effect on reef ecology, and perhaps on near reef migration of tuna, of the selective removal of some small fish species; the identification of alternative sources of bait; and, the development of more effective means of keeping live bait in good condition.

Sea level rise

In common with a number of Pacific Island nations the Maldives has been identified as a nation most at risk in the event of sea level rises which are projected on the basis of global warming from the "greenhouse effect." As an archipelago of low lying sand islands the Maldives would be greatly affected by a rise in mean sea level.[9] Significant sea level rise will change the local dynamics on reefs with possible changes in the locations at which wave energy is reduced and sand or rubble are deposited on the reef top. The actual consequences will depend on the rate of any rise, and the extent to which there may be an increase the incidence of storms.

The areas most likely to be at early risk are those where reclamation has extended the area of islands substantially beyond the area originally covered by natural island formation. The reclaimed edges of such islands are already subject to increased wave energy because they extend beyond the point of depositional energy loss. Increased sea level and consequent wave height relative to the reef edge will further increase the wave energy reaching the reclaimed area.

Problems may be compounded in areas where coral and coral rock collection have reduced the level of the reef surface and the availability of materials to dissipate increased wave energies. A key short-term factor will be the availability of materials for the construction of breakwaters and sea walls. A key to longer term stabilization will be the extent to which reef upper levels can support and maintain active coral growth to nourish the island growth process with new skeletal material.

PROPOSED MANAGEMENT APPROACH

The outcome of the UNESCO mission was a report recommending:

1. The establishment of an executive and consultative system to coordinate actions with respect to conservation and management of the environment and natural resource base of the Maldives.
2. Priority research to define environment management issues and identify possible solutions to known management problems
3. Development of zoning and other measures to provide a range of degrees of access and resource use to meet criteria of reasonable use consistent with conservation

Environment Council

In 1985 by the time of the follow-up mission an Environment Council had been established under the chairmanship of the Minister of Home Affairs. It provided a consultive framework but not, at that time, executive capacity to undertake active coordination to arrange and supervise applied research or to prepare plans.

Research

Projects recommended were:

1. Seismic survey of atoll floor sand deposits to identify the availability of coralline sand for reclamation and building materials.
2. Physical Mapping using LANDSAT data to produce rectified maps at 1:150,000 with indicative bathymetry and shallow water habitat classification.
3. Studies of tidal and wind generated currents to enable planning on the basis of known linkages between probable sites of pollution and prime sources of recruits of reef species.
4. Feasibility of alternative building materials to develop simple, locally applicable, technology to produce building materials of high quality and thus remove the need to harvest reef top corals and reef rock
5. Setting up a code of practice for development and building on and around islands to assemble environmental engineering standards for the design and construction of facilities, to educate developers, designers, builders and inspectors, and thus simplify procedures and increase standards
6. Review of studies of Maldivian marine life to review published and other available data on the marine flora and fauna of the Maldives and thus enable better planning of future research priorities
7. Establishment of baseline sites at Ari Atoll to establish baselines against which the effects of tourist development may be assessed against known prior conditions
8. Feasibility of aquaculture development to determine the possibilities of the culture of marine species becoming a sustainable economic activity to support increasing human population
9. Study of traditional knowledge and management of the marine environment to record traditional understanding and practice, particularly of old people whose knowledge was learned before modern methods and economic pressures affected the traditional relationship between the human population and the reef and island resources
10. Feasibility of compaction and deep-water dumping as means for garbage disposal to develop alternatives to unsightly and potentially unhealthy practices of island or shallow sea disposal of mixed garbage

Zoning and other regulatory measures

As with the other examples presented here, it was apparent that the establishment of an overall environmental plan would have great potential benefits for government and economic development as well as for the retention of key environmental areas and processes. A scheme based on zoning and permits was developed for discussion using Male and Ari atolls as examples (Figures 15.6 and 15.7).

The scheme was based on a gradation from a broad spectrum of use to no human impact. At one extreme is a development zone in which all activities are allowed but those potentially damaging to environmental structure, process or amenity are subject to permit. At the other extreme is a strict nature reserve zone as a reference site in which all direct impacts on the structure and processes are precluded and that may be entered only with a permit for scientific research that cannot be conducted elsewhere. The scheme is summarized in Table 15.1.

The broad concept of zone-based management was accepted, but the resources required to proceed immediately to the long-term commitment of planning, implementation, monitoring, and review were not readily available in the face of other immediate social and economic issues. Several of the identified research priorities, fish and bait fish stock issues, and mapping and resource assessment studies using satellite remote sensing have been addressed with international assistance. There has also been considerable international interest and investment in sea level studies and projections since with its low sandy islands the Republic of Maldives appears to be one of the nations most vulnerable to sea level rise.[11]

The development of a more effective and integrated approach to marine environment management will probably depend on acceptance by the government, and by international aid and financial support agencies, that environmental matters are integral components of virtually all social and economic decisions. Without such acceptance it will take many years before government departments have sufficient numbers of environmentally trained Maldivian professional staff able to work within the decision-making system in such a way that environmental consequences of decisions are anticipated, avoided or addressed before they become problems.

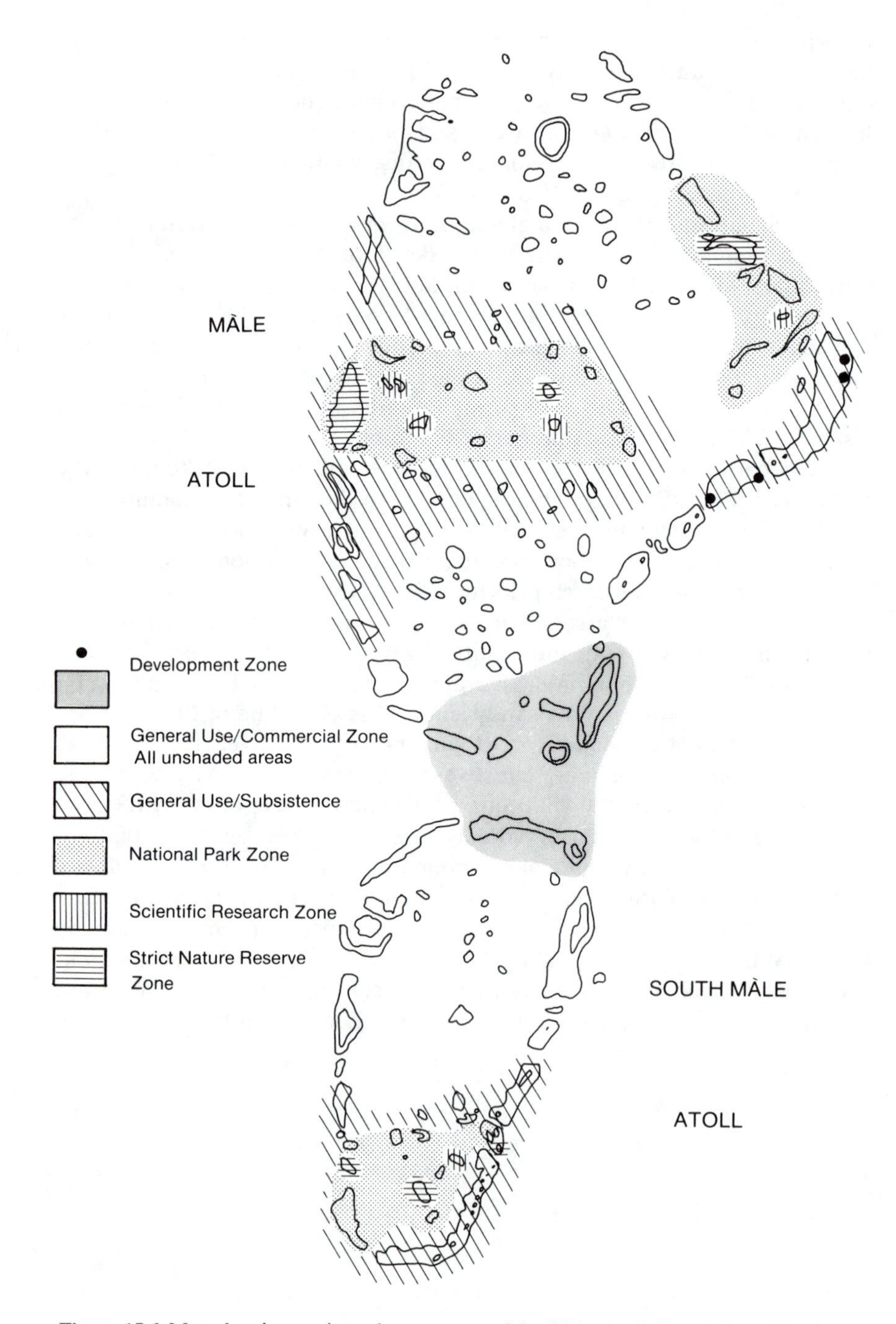

Figure 15.6 Map showing zoning scheme proposed for Male Atoll, Republic of Maldives

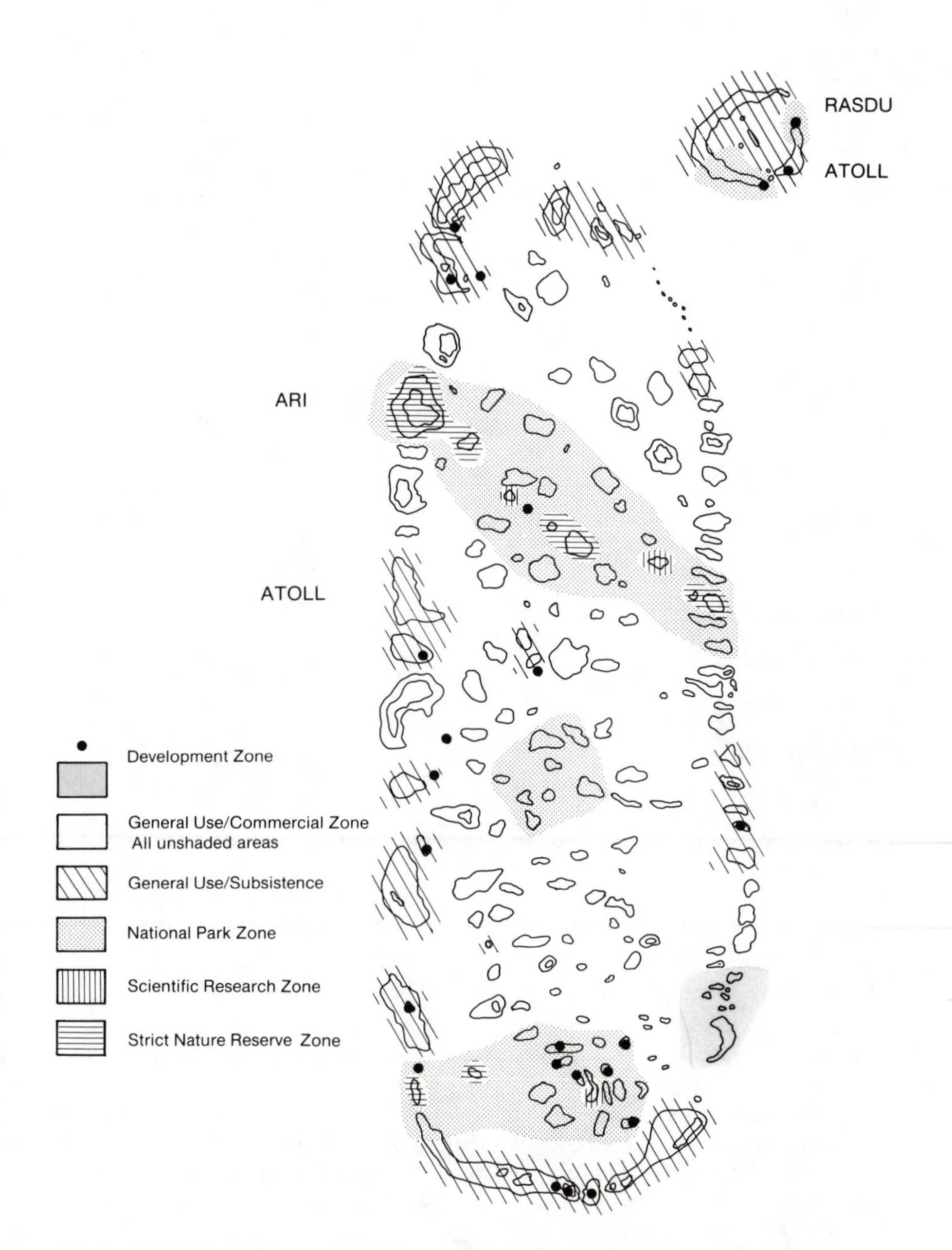

Figure 15.7 Map showing zoning scheme proposed for Ari Atoll, Republic of Maldives

Table 15.1
Conceptual Zoning Scheme for the Maldives

Activity	Devt. GU	Comm GU	Subs	Nat Pk	Sci Res	Nat Res
Reclamation	P	P	P	X	X	X
Building	P	P	P	P	P1	X
Drainage and sewerage	P	P	P	P	P1	X
Dumping/waste disposal	P	P	P	X	X	X
Industrial devt.	P	X	X	X	X	X
Commerce devt.	P	P	X	X	X	X
Major prot./hbr.	P	P	X	X	X	X
Urban devt.	P	P	X	X	X	X
Coral mining	P	P	P	X	X	X
Tourist devt.	P	P	P	X	X	X
Tourist activity	P	P	P	P	X	X
Mariculture	P	P	P	X	P1	X
Pelagic subsist.	P	P	/	X	X	X
Pelagic "exports"	/	/	X	X	X	X
Reef fish subsist.	/	/	/	X	X	X
Reef fish "export"	/	/	X	X	X	X
Shell/coral colln.	/	/	/	X	X	X
Trawling	/	/	X	X	X	X
Other netting	P	P	P	X	X	X
Acquarium fish	P	P	X	X	X	X
Boating	P	P	/	/	X	X
Dive/snorkel	/	/	/	/	X	X
Scientific resch.	/	/	/	/	P	P2

/ Allowed as right
P Allowed subject to permit
P1 Permitted only in connection with approved research
P2 Permitted only for nonmanipulative research that cannot be conducted elsewhere
X Not allowed

Chapter 16

International Arrangements

The need for management to protect marine environments may be particularly apparent in coastal areas of the world that have high human populations with the consequent problems of pollution, and of disputes about allocation and overexploitation of resources. Often the situation is complicated by the interaction of the jurisdiction of several nations. Marine environment management will often involve addressing situations where the environmental costs and impacts experienced in the marine jurisdiction of one nation arise from distant activities in another nation that derives substantial economic benefits from those activities. The complexity of jurisdiction and the extent of the linkages in marine environments and economic activities has led to the formations of many international organizations covering three broad categories of collaboration; maritime law, navigation and shipping, fisheries coordination, and marine sciences.[1] The social and political complexities are so great that an immediate approach toward generally coordinated management is unlikely to be productive. There are means to address international management problems, but to develop and apply them takes time and there is often reluctance to do so until an obvious problem is severe and widespread.

International organizations work through the development of conventions and protocols that are then submitted to governments that may ratify them as formal treaties. If they do so they commit themselves to implement the provisions of the convention or protocol through their own legislation within their own area of jurisdiction. They also commit their residents, citizens, and registered vessels and aircraft to abide by the equivalent legislation of other parties to the convention or protocol.

The process of negotiating international conventions or protocols and having them considered and ratified by the governments of participating nations is usually time-consuming. In many nations ratification may be politically contentious and may be portrayed as ceding influence to an international body and limiting freedom to legislate within areas of national sovereignty. In federal nations such as Australia and the United States, the sensitivity may be particularly acute since a matter covered by an international convention is a matter for the federal government which is responsible for foreign affairs powers. An activity that was previously constitutionally the responsibility of the state level of government becomes a potential subject for federal intervention.

The mechanisms of international procedures may be cumbersome but they are the only means of dealing with matters on the high seas beyond the territorial waters of Exclusive Economic Zone of any nation. They are also often the most acceptable means for developing consistent and even collaborative approaches between nations that do not normally communicate effectively.[1]

THE INTERNATIONAL MARITIME ORGANIZATION (IMO)

An early priority of international management of marine environments was to address the problems of pollution arising from the carriage of increasing quantities of increasingly diverse and increasingly toxic materials in ships. In 1959 the United Nations established the International Maritime Co-ordinating Organization (IMCO)[2] with the following main objectives:

1. To provide machinery for co-operation among governments in the field of governmental regulations and practices relating to the technical matters of all kinds affecting shipping engaged in international trade, to encourage the general adoption of the highest practicable standards in matters concerning maritime safety, efficiency of navigation and prevention and control of marine pollution from ships, and to deal with legal matters related to the purposes of the Organization.
2. To provide for the consideration by the Organization of any matters concerning shipping and the effect of shipping on the marine environment that may be referred to it by any organ or specialized agency of the United Nations
3. To provide for the exchange of information among Governments on matters under consideration by the Organization

Major catastrophes such as the wrecking of the *Torrey Canyon* off Southwest England (1967) and the *Amoco Cadiz*, off the west coast of France (1978) provided added impetus for measures to minimize the likelihood of recurrence of such accidents and to establish effective arrangements for rapid response when they did happen.

Over the years of its operations, IMCO, renamed the International Maritime Organization (IMO) in 1982,[3] has developed a range of conventions and protocols that, if universally implemented, would virtually eliminate operational pollution from ships and should bring about a significant reduction in accidental pollution. These include:

- The International Convention for the Prevention of Pollution of the Sea by Oil (1971) (amended from an earlier convention of 1954)
- The International Convention relating to Intervention on the High Seas in Cases of Oil Pollution Damage (1969)
- The International Convention on Civil Liability for Oil Pollution Casualties (1969)
- The International Convention on the Establishment of an International Fund for Compensation for Oil Damage (1971)
- The Protocol relating to Intervention on the High Seas in cases of Marine Pollution by Substances other than Oil (1969)
- The Convention on the Prevention of Marine Pollution by Dumping Wastes and Other Matters (1972)
- The International Convention for the Prevention of Marine Pollution from Ships (1973 amended 1978)—known as MARPOL 73/78

The listing of the IMO conventions and protocols illustrates how concern has broadened from a focus on oil to a systematic approach to pollution reduction. They started with a concentration on accident prevention and postaccident clean-up issues. They moved on to consider pollution arising from deliberate use of the seas for dumping of waste materials generated on land. The development of MARPOL 73/78 has seen an evolution through the adoption of annexes covering specific forms of operational pollution:

- Annex 1 Oil
- Annex 2 Noxious liquid substances carried in bulk (e.g. chemicals)
- Annex 3 Harmful substances carried in packages (e.g. tanks and containers)
- Annex 4 Sewage
- Annex 5 Garbage

A matter of importance to the members of IMO is the maintenance of the freedom of the high seas and protection of the right of innocent passage for vessels traveling through the territorial waters and other maritime jurisdiction of other nations. There is consequently a preference for measures that can be applied generally as routine operational procedures to control pollution and considerable caution over measures that relate to specific geographic areas. The initial Convention for Prevention of Pollution of the Sea by Oil applied a restriction on polluting activities such as the flushing of oil tanks and discharge of oily bilge water in the territorial sea of nations and established a procedure for applying the same restrictions to other sensitive areas, such as the nonterritorial waters of the Mediterranean Sea and the Great Barrier Reef.

Subsequently, the Dumping Convention of 1972 developed and applied procedures for defining areas in which the dumping of wastes may be allowed.

At the other end of the spectrum of sensitivity and risk procedures, the 1954 International Convention for the Prevention of Pollution of the Sea by Oil established "prohibited zones" of at least 50 miles from the nearest land in which the discharge of oil or of mixtures containing more than 100 parts of oil per million was not allowed. MARPOL 73/78 provided the power to impose strict operational conditions, including exclusion from sensitive areas, on vessels carrying hazardous cargoes. They also include the ability for IMO to agree to declare particularly sensitive sites as "areas to be avoided." It is not illegal for vessels to enter those areas but to do so generally invalidates their insurance against loss, damage, or liability.

The procedures for designating dumping areas, presumably areas of low environmental sensitivity, the prohibited zones and the arrangement for designating areas to be avoided, presumably area of high sensitivity and risk, have established a form of environmental zoning for control of the activities of shipping. The member states of IMO continue to apply careful scrutiny to proposals to limit the freedom of the high seas or the right of innocent passage, but there is likely to be a gradual increase in the number of restricted areas as environmentally sensitive sites are identified, proposed, and adopted. Such sites include areas such as the Galapagos Islands where an accident could have disastrous consequences for endemic fauna of the marine environment. They may also include calving grounds for whales or spawning sites for species of particular ecological or commercial significance.

THE UNITED NATIONS ENVIRONMENT PROGRAM REGIONAL SEAS PROGRAM

The UN Environment Program's Regional Seas Program is concerned with the environmental problems of the oceans on a regional basis.[4] There has been a major emphasis on protection of marine living resources from pollution, but the program has also considered other factors, especially preserving important habitats, managing coastal ecosystems, and protecting coastal soils. Ten regional seas areas have action plans that are in operation or under development:

Mediterranean—adopted 1975
Kuwait—adopted 1978
Caribbean—adopted 1981
West and Central African—adopted 1981
East African—in preparation
East Asian—adopted 1981
Red Sea Gulf of Aden—adopted 1982
South West Pacific—adopted 1982
South East Pacific—adopted 1981
South West Atlantic—in preparation

The program works through the development and adoption of a plan that is then implemented through international treaties. The process can be time consuming and complex once it involves collaboration on measures outside the areas under the certain, three-mile territorial sea jurisdiction of the contracting nations.

The scope and potential of the UNEP Regional Seas Program can be appreciated from the Mediterranean Action Plan. Each of the regional seas has its particular issues, but the enclosed Mediterranean Sea is generally subject to much greater pressures than most other marine areas.

The Mediterranean is an enclosed sea with a small connection through the Straits of Gibraltar to the Atlantic Ocean. It is linked through the Bosporus to another enclosed body of water, the Black Sea. Most of the surrounding nations, many of them heavily industrialized, have large human populations, particularly on the coastal fringe. Many, particularly to the north of the Mediterranean, have a high level of economic development generating wastes that may find their way into the sea. For many of the nations the use of marine resources was not a major element of the national economy until, in the past three decades, coastal leisure and tourism became increasingly important economic activities. Despite the continuing histories of dispute and warfare between neighboring nations in several areas of its coast, the Mediterranean was the first of the UN Environment Program Regional Seas Areas to establish an action plan.

The Mediterranean action plan was developed at a meeting attended by 16 of the 18 Mediterranean coastal nations in early 1975. The legal framework was adopted one year later at a Conference of Plenipotentiaries of the Coastal States of the Mediterranean Region for the Protection of the Mediterranean Seas. It consisted of three instruments:

- The Convention for the Protection of the Mediterranean Sea against Pollution
- The Protocol for the Prevention of Pollution of the Mediterranean Sea by Dumping from Ships and Aircraft
- The Protocol concerning Co-operation in Combating Pollution of the Mediterranean Sea by Oil and other Harmful Substances in Cases of Emergency

The convention is a broadly based document setting out principles and the nations at the conference agreed that no state could be a contracting party to the convention without also becoming a party to at least one of the protocols. The convention and the two protocols came into force in early 1978, and by 1981 they had been ratified by 15 Mediterranean states and the European Economic Community.

In 1980 a further Conference of Plenipotentiaries was held at which a third protocol was adopted to provide for protection of the Mediterranean Sea against pollution from land-based sources. An additional element was added in 1982 with the adoption of a fourth protocol concerning Mediterranean Specially Protected Areas. This was the first occasion where the framework of arrangements under a UNEP action plan was extended beyond the primary process-related concern of pollution to the structural issues of establishment and management of a representative network of protected areas.

For many of the nations, the use of marine resources was initially a matter of local or regional rather than national significance. The principal focus in the initial stages was generated by concern over the impacts of marine pollution upon human health through the absorption or accumulation of toxic materials and human pathogens by seafood. This was augmented by concerns over direct effects of exposure to high levels of human pathogenic organisms during swimming and other water sports in areas subject to the disposal of untreated or partially treated sewage. Later, amenity aspects of pollution joined the human safety aspects as concern arose regarding reduction of the famed transparency or clarity of Mediterranean waters. This was seen as a direct result of an increasing load of fine particulate matter from urban, industrial, and agricultural wastes and an indirect effect through increased phytoplankton levels arising in response to high nutrient levels in sewage wastes.

Changes observed in the structure and species composition of communities of plants and animals of the seabed and water column attracted increasing concern. There was considerable debate over the extent to which they arose from water quality issues as opposed to heavy use for fishing.

A consequence of this concern was recognition of the need for areas protected from fishing and collecting as refuges for sample populations of heavily used species and as recreational areas equivalent to terrestrial national parks providing opportunities for appreciation of nature free from the effects of fishing and collecting.

The process of focusing on amenity and philosophical or aesthetic aspects of environmental quality has expanded as recreation and tourism have grown. Tourism grew from 24 million to 65 million annually between 1960 and 1971 and had probably exceeded 100 million by 1982.[5]

The principal focus of the Mediterranean action plan still remains that of pollution control, but the interaction of affluence, leisure, and tourism have generated a socioeconomic climate in which the costs of preventing pollution, of restoring damaged areas, and of establishing and maintaining protected areas, are increasingly accepted as part of reasonable public and international administration.

The Regional Seas Program is likely to become a major focus for international marine environment management initiatives as protected areas and other approaches develop further in response to recommendations of the World Commission on Sustainable Development, the World Wilderness Congress and the IUCN.[6]

THE BIOSPHERE RESERVE PROGRAM OF UNESCO

The biosphere reserve concept is a radical departure from many recent conservation approaches. It seeks to promote management regimes based on long term understanding of ecosystems and the concept that humans are an integral component of the natural system. The reserve scheme has grown from the Man and Biosphere (MAB) program of UNESCO.

The basis of the biosphere reserve is stewardship providing for sustainable development through zoning plans in which different areas serve the roles of conservation—preservation of species and habitats; logistics—providing controlled locations for research into ecological systems and human interactions with them; and development—controlled, sustainable use.[7] Several hundred biosphere reserves have been declared, linking diverse representative ecosystems in a global network of conservation, research and monitoring, and public education. Nevertheless, by 1989 no truly marine biosphere reserves had been proclaimed.

The biosphere reserve program was formalised in 1971. In the early years, the relative ease with which the scheme could be advanced by superimposing biosphere reserves upon existing national parks meant that most were superfluous from a management standpoint although they had great symbolic value.[8] The program gained strength and started to move away from its largely symbolic role but the conservation role of the reserves was often emphasised at the expense of the logistic and development roles.[9]

To meet the objectives, the program should help to preserve representative examples of ecosystems, and aid in the establishment of centers for monitoring, research, education and training. It should also provide a framework within which government decision-makers, scientists, local people and managers can cooperate effectively in managing natural resources.[10] There are good examples on land where people with different interests have come together in the biosphere reserve forum to work out wise solutions for resource use issues.[11] This philosophy is applicable to the scale and dynamics of marine environments where in crowded, closely linked waters an even greater need can exists for such forum building mechanisms.

Nevertheless, the nature and scale of marine systems and the conventions of marine law are such that some reinterpretation of biosphere reserve criteria may be needed for the purposes of marine areas.

On land, the essential design of a biosphere reserve is centered upon one or more core areas providing a self-sufficient *in situ* conservation unit, representing a defined ecotone, that can be strictly protected in the long term. Around the core areas are buffer zones in which conforming use of the resources of the environment is controlled to prevent impacts which are likely to damage the core.

In the sea, a self-sufficient *in situ* conservation unit for an ecotone with planktonic larval distribution may be very large, making the concept of a substantial strictly protected core impossibly large.

On land, the primary issue may be one of species or habitat protection which is addressed by the conservation role of the biosphere reserve. In the sea, the primary issues may be matters of the logistic difficulty of maintaining substantial research or monitoring programs or the development concern of sustaining the resource base of coastal communities.

With these management priorities the core of reserves in the sea may be the establishment of a viable logistic network of research or reference sites. Batisse[12] commented on the lack of attention to the logistic role of biosphere reserves on the early history of the program. This role appears to be particularly appropriate in marine environments where linkages often extend over two or more systems of human jurisdiction. Virtually any site is likely to be directly or indirectly affected by human activity so an extensive and active global network of marine reference sites could be an important contribution to the ability to study human and other impacts on marine environments and processes over long distances. Many such reference sites need not be managed as strict nature reserves. Provided they are managed in such a way that the research value of the sites is not compromised, compatible uses—particularly non-consumptive or non-extractive activities may be encouraged.

Perhaps the greatest potential contribution of marine biosphere reserves may be to provide additional means for nations that share, or jointly impact upon, marine resources to develop management regimes to sustain them despite social, economic of political differences. If this is to be the case there is a need to develop a theoretical and operational basis for marine biosphere reserve planning.

Chapter 17

Prospects for Progress in Marine Environment Management

The most pressing marine management issue is the use of rivers and the sea for the disposal of waste by-products from terrestrial activity. Traditionally this has been the easy and cheap solution. However, as the limitations and environmental costs of that solution become increasingly apparent, human communities must turn to methods that incur economic costs in order to remove or reduce environmental costs. Solving the problems of poor practices of urban and industrial waste disposal and of poor agricultural use of soils and chemicals will involve a radically new approach to valuing and costing the use of natural waters.

Maintenance of coastal and marine environments depends increasingly on paying the costs of land use practices and treatment of wastes to ensure that discharge levels of nutrients, silt, heavy metals, and synthetic molecules are safely within sustainable thresholds. As with atmospheric environmental pollution control, this should involve international collaboration because, in a competitive market, prices are set by the lowest common denominator. If that lowest common denominator is a producer who does not pay the cost of pollution prevention, short-term economic logic will often overcome long-term environmental and economic logic.

In the case of over exploitation of biological resources the tragedy of the commons is having its impact in the evolution of national and international fishing policies. Increasingly, management for sustainable fisheries involves restricting and policing access to fish stocks. This in turn will involve allocating access to stocks between fishery components, subsistence, recreation, local market, and industrial export.

In the case of alienation by reclamation of productive estuarine, coastal, and shallow marine habitat, effective management will increasingly involve accounting for the value of those environments to coastal processes, fisheries and recreation in order to assess the enduring costs imposed by the apparently cheap option of "creating" land by alienation of intertidal or shallow marine environments.

For the foreseeable future the major management priorities are to develop practical skills in marine environment planning and management and to generate widespread community awareness of the problems and costs of inadequate marine environment management

RECREATION AND TOURISM AS PRIMARY USERS OF MARINE ENVIRONMENTS

For many people in the developed world, automation and affluence have reduced the amount of time spent in employment or business and thus provided the means and the opportunity to take part in leisure activities. Through tourism some of the leisure and recreation needs of affluent and largely urban communities may be fulfilled at distant locations. Provision of opportunities for leisure or recreation activities through tourism can become the means to bring economic activities and benefits from urban centers to more remote areas of developed countries and internationally to developing countries.

The coordination of long-term planning and management for recreation and tourism in environmentally attractive areas is one of the most important challenges of coastal and marine environment management. As a reasonable use based on appreciation and enjoyment of the environment, tourism can provide motivation for conservation and lead influential decision-makers in human communities to appreciate the values of high environmental quality. It can generate long term economic and social benefits locally, nationally, and for the global community.

Coastal and marine environments are increasingly important in the provision of leisure, recreation, and tourism opportunities. For passive recreation the interface between land and open water with fresh air, natural light, and, often, the rhythmical sound of waves meeting the shore provides a soothing contrast to the bustle of city life. Aesthetic or cultural activities based on observation or experience of natural features can require access to substantial natural areas with diverse or unusual biological communities. Physical leisure activities such as hiking, boating, or diving may be heightened by access to substantial areas that are apparently remote and undeveloped.

The forms of leisure activities and their demands for natural resources have changed with new technology and with different expectations flowing from education. Fishing and collecting, still the prime motivation for many recreational visits to coastal and marine areas, have been joined and in some cases displaced by activities in which visitors come to see and appreciate the beauty and learn something of the nature of coastal and marine physical structure and biological communities.

The situation is complicated by amenity issues and changing perceptions of amenity. Robinson[1] describes the situation whereby in a natural or little developed situation most "insiders" or people living in or near a natural environment value it for the economic potential for extraction or development, whereas "outsiders" those living far away, are more likely to value it for aesthetic, philosophical, or cultural attributes.

The implementation of management measures almost inevitably changes the nature of the recreational experience and, with that, the aproach to the natural environment. There is consequently a tendency for the amenity and motivation of tourist and recreational facility development to creep, or change emphasis as development progresses.

The first stage of tourism development typically occurs to capitalize on "outsider" values, providing access to natural attractions. It is likely to introduce relatively small or simple structures or vessels providing an opportunity to conduct recreational activities. The recreational setting is heightened by access to otherwise undisturbed or little disturbed natural landscapes or seascapes perhaps offset by apparently tranquil agricultural or fishing settlements. This first stage often identifies opportunities to realize "insider" values through sale and development of real estate.

In the second stage the initial focus on natural attractions leads to the need for recreational management strategies with visitor management and hardening, to protect sensitive sites and enable them to support more visitors, through works such as the construction of fences, walking tracks, car parks or boat moorings. In the nearby settlements it changes the economic emphasis with the introduction and operation of ancillary infrastructure such as specialist shops, restaurants, bars and sports facilities, which depend on tourism for their viability. This leads to the conversion or development of housing to accommodate visitors and people immigrating to provide goods and services to the tourist industry.

A third stage may follow with urban economic diversification. Particularly in areas with mild climates, erstwhile tourists retire after accumulating wealth in industrial areas. Commerce develops to support a growing population. Light industry is sought to diversify the employment and economic base. Tourism develops through construction and promotion of synthetic urban facilities such as golf courses, country clubs, and theme parks, which may have little connection with the natural environment beyond fresh air, sunlight, and, perhaps, sea and sand. Other facilities such as casinos, nightclubs, shopping and entertainment centers may create their own entirely synthetic environment. The economic attractiveness of such developments may appear to depend on access to cheap land and this may lead to pressure to alienate coastal environments such as mangroves,

results generally indicate the adoption of plans without adequate commitment to a mechanism to ensure their implementation. Although particularly acute in developing nations, the same problems were reported from more developed nations. "Paper parks" generate the illusion, and often the political belief, that the necessary steps have been taken while in reality nothing has changed.

The challenge is to build into initial planning a realistic basis for perpetual support of adequate management. That basis should be tied to economic developments that use or have impacts upon the resources of the managed area. It should also include a "start-up" component to cover a period of five years or so while staff are recruited and trained, equipment acquired and a stable sustainable operational basis established. The priority tasks of the initial period also include public education and, possibly, economic adjustment so that the local communities and users are generally motivated to support the objects of the resource management regime.

In part this challenge should be met if planning involves and reflects the ecological and economic components from the start. In part it requires a sustained educational program targeted particularly at socioeconomic decision makers. Such a program should focus on the inevitable nexus of economic and ecological factors so that it becomes a fundamental accepted factor in social and economic planning underpinning the more usual concerns over growth, employment, security and community development. In many ways the most important educational targets are economic planners, national and international development agencies, bankers, and investment advisors. In the developed nations the key is to convince government of the long-term importance of sustainable environmental management even if, on occasion, it conflicts with short-term convenience.

In the case of developing countries such as Ecuador and the Maldives, governments may recognize the importance in the medium and long term of conservation planning and management but face even more acute problems in allocating adequate resources of national funds, or aid funds, in the face of strong competition from pressing immediate socioeconomic problems. In such circumstances, establishment of a viable conservation scheme will probably not occur without international initiatives to build environmental planning into aid and development projects and to provide commitments of aid, expertise, and training for a sufficient period to establish a viable management unit. The key to viability is sustainable funding linked to the extent of social and economic activities that have impacts on the area in question.

The role of international agencies and agreements is critical. In many cases they are essential because of the scale of the systems involved. In the overall context they are important in setting an attitudinal climate that will flow into the general decision-making framework of global economic policy. For the developing nations they are critical because many of those nations have very large areas of marine jurisdiction and very limited management resources. The provision of substantial technical support and training is crucial to the implementation of management of the marine environments that support the food supply and economies of many of the world's least developed nations.

There is a well-developed network for training and technical support in the areas of marine science. UNESCO, through its Division of Marine Science, supports the International Oceanographic Commission (IOC) with regional committees for each of the world's major ocean areas. IOC and other international specialist organisations such as the International Association for Biological Oceanography and the Scientific Committee for Oceanic Research support and promote multilateral and regional collaborative programs of marine science research and training.

Marine environment management has developed under the auspices of the marine science arrangements, but, as yet, arrangements for international collaboration on training and implementation of management generally are not as well developed as those for research and monitoring. The pressing need, particularly in many developing countries is for marine management training as opposed to marine science training.

Management training should recognize and accept the critical specialist role of marine scientists as primary sources of information and as designers and implementers of monitoring programs. Accepting this, marine area management training should introduce managers to the key issues of marine ecology that affect the managed area, incorporate social and economic factors, and provide a solid grounding in the routine but vital tasks of day-to-day management. The most important link in the management chain is generally the local field manager, warden, ranger, patrol officer or boatman. Effective training for such people can often only be carried out effectively in their home situation with materials and examples that reflect local language and local examples. The most critical elements are often the basic management issues of patrol organization, public contact, and equipment maintenance, without which the management plan is unlikely to work.

The planned collaboration of UNEP and IUCN to build upon the basis of the UNEP Regional Seas Program to strengthen management should lead to regional collaboration, particularly in the area of management training. A program by East Asian nations to develop training materials under the UNEP Co-ordinating Body for East Asian Seas (COBSEA) is an example of such collaboration. Development of more advanced courses and a network to enable more senior managers to share experiences and collaborate in regional problem solving is a logical next step. This process may be enhanced through the involvement of UNESCO through the biosphere reserve program extended to apply to marine environments. The key will be to achieve continuing collaboration between the participants in the international initiatives.

Marine management has been forced to develop in great haste. The intention of this book has been to introduce and discuss some of the key issues. The hope is that it will play a part in developing a professional body of marine environment managers to complement and implement the findings of marine science, and so provide for sustainable development and conservation of the seas and coastal fringes of the world.

Notes

Introduction

1. World Commission on Environment and Development. 1978 *Our Common Future* (Oxford: Oxford Univ. Press), 400 pp.; E. M. Borghese. 1986. *The Future of the Oceans—A Report to the Club of Rome* (Montreal: Harvest House), 144 pp.

2. R.V. Salm and J. R. Clark. 1984. *Marine and Coastal Protected Areas: A Guide for Planners and Managers* (Gland, Switzerland: IUCN), 302 pp.

3. R.A. Kenchington and B.E.T. Hudson, eds. 1987. *UNESCO Coral Reef Management Handbook, 2nd ed.* (Jakarata: UNESCO ROSTSEA), 321 pp.

4. Ibid.

Chapter 1

1. B. Meehan. 1982. *Shell Bed to Midden* (Canberra: Aust. Inst. of Aboriginal Studies), 189 pp.; I. Cornwell. 1964. *The World of Ancient Man* (London: Phoenix); J.G.D. Clark, 1952. *Prehistoric Europe. The Economic Basis* (London: Methuen).

2. B. Landstrom. 1961. *The Ship: A Survey of the History of the Ship from the Primitive Raft to the Nuclear Powered Submarine* (trans. M. Phillipps) (London: Allen and Unwin), 309 pp.; various authors. 1980, 1981. *The Ship* (London: HMSO). 10 vol., 627 pp.

3. Ibid.

4. H.N. Moseley. 1892. *Notes by a Naturalist on the Challenger* (London: J. Murray), 540 pp.

5. D.H. Cushing. 1988. *The Provident Sea* (Cambridge: Cambridge Univ. Press), 329 pp.

6. *Halsbury's Laws of England*. 1954. Vol 1. *Admiralty* (London: Butterworth), 208 pp.

7. R. E. Johnannes. 1981. *Words of the Lagoon; Fishing and Marine Lore in Palau district of Micronesia* (Berkeley: Univ. of California Press); M. Titcomb. 1972. *Native Use of Fish in Hawaii* (Honolulu: Univ. of Hawaii Press).

8. P.M. Fye. 1974. "The Future," in *Oceanography, the Last Frontier*, ed. R.C. Vetter (Washington, DC: Voice of America, Forum Series), pp. 419–30.

9. J.A. Crutchfield. 1982. Keynote address in *Policy and Practice in Fisheries Management*, eds. N.H. Sturgess and T.F. Meany, Proc. National Fisheries Seminar 1980, Melbourne, Autralian Govt. Publ. Serv., pp. 1–38.

10. T.G. Kailis. 1982. "Limited Entry—An Industry View", in *Policy and Practice in Fisheries Management*, pp. 77–86.

11. D. Cormack. 1983. *Response to Oil and Chemical Marine Pollution* (London: Elsevier), 530 pp.; M. Ruivo, ed. 1972. *Marine Pollution and Sea Life* (London: Fishing News Books), 624 pp.

12. R. Carson. 1962. *Silent Spring* (Boston: Houghton Mifflin), 368 pp.; B. Commoner. 1967. *Science and Survival* (London: Gollancz), 122 pp.

13. T. Heyerdahl. 1971. *The Ra Expeditions*, trans. from the Norwegian by Patricia Crampton (London: Allen and Unwin), 334 pp.

14. S.A. Gerlach. 1981. *Marine Pollution: Diagnosis and Therapy*, trans. R. Youngblood and S. Messele-Weiser (New York: Springer Verlag), 218 pp.

15. S. Mankaby, ed. 1984. *The International Maritime Organisation* (London: Croom Helm), 376 pp.

16. IUCN. 1976. Proceedings of an Inernational Conference on Marine Parks and Reserves, Tokyo, 12–14 May 1975, IUCN Publications New Series 37.

17. M.E. Silva, E.M. Gately, and I. Desilvestre. 1986. *A Bibliographic Listing of Coastal and Marine Protected Areas: A Global Survey*, (Woods Hole, MA: Woods Hole Oceanogr. Inst. Tech). Report. WHOI-86-11.

18. Ibid.

19. R.V. Salm, S. Sudargo, and E. Mashudi. 1982. "Marine Conservation in Indonesia," *Parks* 4(1): 1–6.

20. R.A. Rahman and M. Ibrahim, eds. 1987. *Proposed Mangement Plan for the Pulau Redang Marine Park* (Selangor: Universiti Pertanian), 336 pp.

21. G.G. Kelleher and R.A. Kenchington. 1982. "Australia's Great Barrier Reef Marine Park: Making Development Compatible with Conservation," *Ambio* 11 (5): 262–67.

22. International Union for the Conservation of Nature and Natural Resources. 1980. *World Conservation Strategy: Living Resources Conservation for Sustainable Development* (Morges, Switzerland: IUCN), 55 pp.

23. World Commission on Environment and Development. 1987 *Our Common Future* (Oxford Univ. Press), 400 pp.

24. G.G. Kelleher and R.A. Kenchington (in press) *Guidelines for Establishing Marine Protected Areas* (Morges, Switzerland: IUCN).

Chapter 2

1. E.D. Brown and R.R. Churchill. eds. 1987 *The UN Convention of the Law of the Sea: Impact and Implementation*, Proc. Law of the Sea: Inst. 19th Annual Conference, Honolulu, The Law of the Sea Institute, Univ. of Hawaii), 639 pp.

2. T. Beer. 1983. *Environmental Oceanography* (New York: Pergamon), 262 pp.

3. A.N. Strahler. 1973. *Introduction to Physical Geography*, 3rd ed. (New York: Johyn Wiley), 468 pp.

4. R.D. Lumb. 1978. *The Law of the Sea and Australian Off-Shore Areas*, 2nd ed. (Brisbane, Australia: Univ. of Queensland Press), 205 pp.

5. L.M. Alexander. 1987. "The Indentification of Technical Issues of Maritime Boundary Delimitation within the Law of the Sea Convention Context", in *The UN Convention on the Law of the Sea*, pp. 272–87; J.R.V. Prescott. 1987. "Straight Baselines: Theory and Practice", in *The UN Convention on the Law of the Sea*, pp. 288–318; P.B. Beasley, 1987. "Maritime Boundaries: A Geographical and Technical Perspective", in *The UN Convention on the Law of the Sea*, pp. 319–39.

6. P. Hulm. 1983. *A Strategy for the Seas: The Regional Seas Programme, Past and Future* (Geneva: UNEP), 28 pp.

Chapter 3

1. R.S.K. Barnes and R.N. Hughes. 1988. *An Introduction to Marine Ecology*, 2nd ed. (Oxford: Blackwell Scientific Publications), 351 pp.

2. R.S. Scheltema. 1989. "On the Children of Benthic Invertebrates: Their Ramblings and Migrations in Time and Space", in *Environmental Quality and Ecosystem Stability*, eds. E. Spanier, Y. Steinberger, and M. Luria (Jerusalem: ISEEQS Publication), pp. 93–112.

3. M.B. Schaefer. 1970. "Men, Birds and Anchovies in the Peru Current—Dynamic Interactions", *Trans. Amer. Fisheries Soc.* 99 (3): 461–467.

4. W. Heape and F.H.A. Marshall. 1969. *Emigration, Migration and Nomadism* (New York: Kraus), 369 pp.

5. Open University Oceanography Course team. 1987. *Seawater: Its Composition, Properties and Behaviour* (Oxford: Pergamon).

6. Barnes and Hughes. 1988. *An Introduction.*

7. R.E. Thresher. 1984. *Reproduction in Reef Fishes* (Neptune City, NJ: TFH Publications), 399 pp.; S.E. Stanwych, ed. 1979. *Reproductive Ecology of Invertebrates* (Columbia: Univ. of South Carolina Press), 283 pp.

8. D.R. Stoddart and R.D. Johannes, eds. 1978. *Coral Reefs; Research Methods* (Paris: UNESCO), 581 pp.

9. D.B. Ivan R. Claasen and P.A. Pirrazzoli. 1984. "Remote Sensing: A Tool for Management," in *UNESCO Coral Reef Mangement Handbook*, eds. R.A. Kenchington and B.E.T. Hudson (Jakarta: UNESCO ROSTSEA), pp. 68–88; C.J. Johannsen and J.L. Sanders, eds. 1982. *Remote Sensing for Resource Management* (Ankeny, IO: Soil Conservation Society of America), 665 pp.

Chapter 4

1. L.S. Hamilton and S.C. Snedaker, eds. 1984. *Handbook for Mangrove Area Managment* (Honolulu: United Nations Environment Program and East-West Center, Environment and Policy Institute), 123 pp.

2. F. Berkes. 1985. "Fishermen and the Tragedy of Commons", *Environmental Conservation* 12 (3):199–206.

3. T. Heyerdahl. 1971. *The Ra Expeditions*, trans. from the Norwegian by Patricia Crampton (London: Allen and Unwin), 334 pp.; A. Carr, 1987. "Impact of Nondegradable Marine Debris on the Ecology and Survival Outlook of Sea Turtles", *Mar. Poll. Bull.* 18(6):353–56.

4. S. Mankaby, ed. 1984. *The International Maritime Organisation* (London: Croom Helm), 376 pp.

5. R.A. Geyer, ed. 1981. *Marine Environmental Pollution*, Vol. 2. *Dumping and Mining* (Amsterdam: Elsevier), 574 pp.

6. P.L. Pearce. 1988. *The Ulysses Factor: Evaluating Visitors in Natural Settings* (New York: Springer-Verlag).

7. P. Moran. 1986. "The Acanthaster Phenomenon", *Oceanogr. Mar. Biol. Ann. Rev.* 24:398–480.

8. R.A. Kenchington. 1987. "Acanthaster Planci and Management of the Great Barrier Reef", *Bull. Mar. Sci.* 41(2):552–60.

9. B. Salvat, ed. 1987. *Human Impacts on Coral Reefs: Facts and Recommendations* (French Polynesia: Antenne Museum E.P.H.E.), 253 pp.

10. Hamilton and Snedaker, eds. 1984. *Handbook for Mangrove Area Management*, 123 pp.

11. Berkes. 1985. "Fishermen and the Tragedy of the Commons".

12. J.J. Munro, J.D. Parrish, and F.H. Talbot, 1987. "The Biological Effects of Intensive Fishing upon Coral Reef Commmunities". in *Human Impacts on Coral Reefs*, pp. 41–50.

13. A.C. Alcala and E.D. Gomez. 1987. "Dynamiting Coral Reefs for Fish: A Resource Destructive Fishing Method", in *Human Impacts on Coral Reefs*, pp. 51–60.

14. L.G. Eldredge. 1987. "Poisons for Fishing on Coral Reefs", in *Human Impacts on Coral Reefs*, pp. 61–66.

15. W.F. Royce, 1984. *Introduction to the Practice of Fishery Science* (Orlando: Academic Press), 428 pp.

16. B. Salvat. 1987. "Dredging in Coral Reefs", in *Human Impacts on Coral Reefs*, pp. 165–84.

17. C. Baldwin, ed. 1988. "Workshop on Nutrients in the Great Barrier Reef Region", Great Barrier Reef marine Park Authority, Workshop Series 10, 191 pp.; R.A. Kenchington and B. Salvat. 1987. "Man's Threat to Coral Reefs", in *UNESCO Coral Reef Managment Handbook*, eds. R.A. Kenchington and B.E.T. Hudson (Jakarta: UNESCO ROSTSEA), pp. 23–28.

18. R.A. Kenchington. 1984. "The Concept of Marine Parks and its Implementation", in *Capricornia Section, Great Barrier Reef*, Roy. Soc. Qd. Symp., pp. 153-58.

19. R.A. Kenchington. 1979. "Marine Park Management Principles", Workshop on the Northern Sector of the Great Barrier Reef, GBRMPA, Townsville, pp. 422–31.

20. J.C. Sorenson, S.T. McCreary, and M.C. Hershman. 1984. "Institutional Ar-

rangements for the Management of Coastal Resources", Renewable Resources Information Series, Coastal Management Publication No. 1, Research Planning Institute, Columbia SC, 165 pp.

21. Kenchington. 1979. "Marine Park Management Principles".

22. T.J. Hundloe. 1987. "Economic Studies", in *UNESCO Coral Reef Management Handbook*, pp. 89–106.; J.A. Sinden and AC. Worrell. 1979. *Unpriced Values: Decisions without Market Prices* (New York: John Wiley), 511 pp.

23. R.A. Kenchington, (in press), "Managing Marine Enviroments", in *Pacific Rim—86*, ed. R. Harris (Canberra)

24. Ibid.

Chapter 5

1. K. Ruddle K. and R.E. Johannes, eds. 1985. *The Traditional Management of Coastal Systems in Asia and the Pacific* (Jakarta: UNESCO ROSTSEA).

2. G.G. Kelleher and R.A. Kenchington. 1982. "Australia's Great Barrier Reef Marine Park: Making Development Compatible with Conservation," *Ambio* 11 (5): 262–67.

3. G. Kelleher and B. Lausche. 1984. "Review of Legislation," in *UNESCO Coral Reef Management Handbook*, (2nd ed., 1987), eds R.A. Kenchington and B.E.T. Hudson (Jakarta: UNESCO), pp.47–51.

4. Kelleher and Kenchington. 1982. "Australia's Great Barrier Reef Marine Park:"

5. International Union for the Conservation of Nature and Natural Resources. 1980. *World Conservation Strategy: Living Resources Conservation for Sustainable Development* (Morges, Switzerland: IUCN).

6. G.E. Machlis and D.L. Tichnell. 1985. *The State of the World's Parks* (Boulder CO: Westview Press), 131 pp.

7. Ibid.

8. Great Britain, Foreign and Commonwealth Office, 1983. Final Act of the International Conference on the Conservation of Wetlands and Waterfowl, Ramsar, Iran, 3 February 1971, and, Convention on Wetlands of International Importance Especially as Waterfowl Habitat, Paris, 12 July 1972, H.M.S.O., London, Parl. Papers by Cmnd. 5483, 19 pp.

Chapter 6

1. R.A. Kenchington and B.E.T. Hudson, eds. 1987. *UNESCO Coral Reef Management Handbook, 2nd ed.* (Jakarta: UNESCO ROSTSEA), 321 pp.

2. R.A. Kenchington. (1990) "Planning the Great Barrier Reef Marine Park," in *Conserving our Marine Heritage*, ed. R. Grahame (Waterloo, Canada: Proc. Symps. Univ.), pp. 35–54.

3. F. Berkes. 1985. "Fishermen and the Tradedy of the Commons," *Environmental Conservation* 12(3):199–206.

4. G. Kelleher and B. Lausche. 1984. "Review of Legislation," in UNESCO Coral Reef Management Handbook, pp. 47-51.

5. W.L. Meyers and R.L. Shelton. 1980. *Survey Methods for Ecosystem Mangement* (New York: John Wiley), 403 pp.; K. Myers, C.R. Margules, and I. Musto, eds. 1984. *Survey Methods for Nature Conservation; Workshop held at Adelaide University, 31 August–2 September, 1983*, Canberra, CSIRO.

Chapter 7

1. A.H. Robinson. 1987. "Staff Traning for Coral Reef and Other Marine Area Management," in *UNESCO Coral Reef Managment Handbook*, eds R.A Kenchington and B.E.T. Hudson: (2nd ed. 1987) (Jakarta:UNESCO), pp. 147–62.

2. B.E.T. Hudson, 1987. "User and Public Education," in *UNESCO Coral Reef Management Handbook*, pp. 163–76.

3. S. McB.. Carson. 1978. *Environmental Education: Principles and Practice* (London: Arnold), 258 pp.

4. G. Claridge. 1987. "Assessing Development Proposals," in *UNESCO Coral Reef Management Handbook*, pp. 131–38.

5. Ibid.

Chapter 8

1. D. Hopley. 1982. *The Geomorphology of the Great Barrier Reef* (New York: John Wiley), 453 pp.

2. W.G.H. Maxwell. 1968. *Atlas of the Great Barrier Reef* (Amsterdam: Elsevier), 258 pp.

3. Hopely. 1982. *The Geomorphology of the Great Barrier Reef.*

4. S. Wells, ed. 1988. *Coral Reefs of the World: Vol. 3 The Western Pacific* (Cambridge: IUCN/UNEP; E.M. Wood. 1983. *Corals of the World* (Neptune city, NJ: T.F.H. Publications), 235 pp.

5. J.E.N. Vernon and M. Pichon. 1982. *Scleractinia of Eastern Australia. Pt IV. Family Poritidae* (Townsville, Old: Aust. Inst. of Marine Science), 159 pp.

6. C. Baldwin, ed. 1988. Workshop on Nutrients in the Great Barrier Reef Region, Great Barrier Reef Marine Park Authority, Workshop Series, 191 pp.

7. W. Saville-Kent. 1893. *The Great Barrier Reef of Australia: Its Products and Potentialities* (London: Allan), 387 pp.

8. T. Hundloe. 1985. Fisheries of the Great Barrier Reef, Great Barrier Reef Marine Park Authority, Special Publication Series 2, 158 pp.

9. R. Claringbould, J. Deakin, and P. Foster. 1984. Data Reviuew of Reef Related Tourism 1946–1980, Great Barrier Reef Marine Park Authority, 119 pp.; S.M. Drim. 1987. Great Barrier Reef Tourism—A Review of Visitor Use, Great Barrier Reef Marine park Authority, 49 pp.

10. F. Gray and L. Zann, eds. 1987 Traditional Knowledge of the Marine Environment in Nothern Australia, Great Barrier Reef Marine Park Authority, Workshop Series No. 8, 196 pp.

11. J.C.H. Foley. 1982. *Reef Pilots—The History of the Queensland coast and Torres Strait Pilot Service* (Sydney, NSW, Australia: Banks and Bros.), 202 pp.

12. J.C. Beaglehole. 1962. *The* Endeavour *Journal of Joseph Banks* 1768—1771 (Syndey, NSW, Australia: Angus and Robertson).

13. R. Hughes, 1987. *The Fatal Shore: A History of the Transportation of Convicts to Australia*, 1787–1868 (London: Collins Harvill); M. Clark. 1962. *A History of Australia*. (Melbourne: Melbourne Univ. Press), 6 vols.

14. Foley, 1982. *Reef Pilots*.

15. W. Saville-Kent, 1983. *The Great Barrier Reef of Australia: Its Products and Potentialities* (London: Allan), 387 pp.

16. C.M. Yonge. 1938. "Origin, Organisation and Scope of the Expedition, Great Barrier Reef Expedition 1928–29 Scientific Reports," *Brit. Mus. Nat. Hist.* 1: 1–11.

17. Beaglehole, 1962. *The* Endeavour *Journal.*

18. O.A. Jones. 1974. *The Great Barrier Reef Committee 1922—1973*, Proc 2nd Internat. Coral Reef Symp. Great Barrier Reef Colmmittee, Brisbane, Qld, Australia, pp. 733–37.

19. J.H. Connell. 1973. "Population Ecology of Reef Building Corals," in *Biology and Geology of Coral Reefs 2, Biology* eds. O.A. Jones and R. Endean, (New York: Academic Press), 205–46.

20. R.G. Pearson and R. Endean. 1969. "A Preliminary Study of the Coral Predator Acanthaster Planci (L.) (Asteroidea) on the Great Barrier Reef," Dept. Harbours and Marine, Brisbane, Qld., Australia, *Fish Notes* 3:27–55.

Chapter 9

1. R. Fitzgerald. 1984. *From 1915 to the Early 1980s: A History of Queensland* (Brisbane: Univ. of Queensland Press), 653 pp.

2. R. Carson. 1962. *Silent Spring* (Boston:Houghton Mifflin), 368 pp.; B. Commoner. 1967. *Science and Survival* (London: Gollancz), 122 pp.

3. P. Clare. 1971. *The Struggle for the Great Barrier Reef* (Melbourne: Collins), 224 pp.

4. J. Wright. 1977. *The Coral Battleground* (Melbourne: Thomas Nelson), 203 pp.

5. Fitzgerald. 1984. *From 1915 to the Early 1980s*.

6. O.A. Jones. 1974. *The Great Barrier Reef Committee 1922—1973*, Proc 2nd Internat Coral Reef Symp. Great Barrier Reef Committee, Brisbane, Qld, Australia, pp. 733-737.

7. D. Cormack. 1983. *Response to Oil and Chemical Marine Pollution* (London: Elsevier), 530 pp.

8. Royal Commissions into Exploratory and Production Drilling for Petroleum in the Area of the Great Barrier Reef. 1974. Report, Australian Government Publishing Service, Canberra, A.C.T. Australia, 2 vols, 1,052 pp.

9. Ibid.

10. D.C. Potts. 1982. "Crown of Thorns Starfish: Man-induced Pests or a Nat-

ural Phenomenon?" in *The Ecology of Pests*, eds. R. Kitching and R.E. Jones (Canberra: CSIRO), pp. 55–86.

11. P. Moran. 1986. "The Acanthaster Phenomenon," *Oceanogr. Mar. Biol. Ann. Rev.* 24:398–480.

12. R.A. Kenchington. 1978. "Crown-of-Thorns Crisis in Australia: A Retrospective Analysis," *Environ. Conserv.* 11 (2):103–18, and 1987. "Acanthaster Planci and Managment of the Great Barrier Reef," *Bull. Mar. Sci.* 41(2):552–60.

13. R.A. Kenchington and R.G. Pearson. 1982. *Crown of Thorns Starfish on the Great Barrier Reef: A Situation Report*, Proc 4th Internat. Coral Reef Symp, 2:597–600.

14. R. Claringbould, J. Deakin, and P. Foster. 1984. Data Review of Reef Related Tourism 1946–1980, Great Barrier Reef Marine Park Authority, 119 pp.

15. Clare, 1971. *Struggle for the Great Barrier Reef.*

Chapter 10

1. Colin Howard. 1978. *Australia's Constitution* (Harmondsworth, U.K. Penguin).

2. D. Hopely. 1982. *The Geomorphology of the Great Barrier Reef* (New York: John Wiley), 453 pp.

3. R.D. Lumb, 1978, The Law of the Sea and Australian Off-Shore Areas, 2nd ed. (Brisbane: Univ. of Queensland Press), 205pp.

Chapter 11

1. Royal Commission into Exploratory and Production Drilling for Petroleum in the Area of the Great Barrier Reef. 1974. Report, Australian Government Publishing Service, Canberra, A.C.T. Australia, 2 vols., 1,052 pp.

2. J. Wright. 1977. *The Coral Battleground* (Melbourne: Thomas Nelson), 203 pp.

3. International Union for the Conservation of Nature and Natural Resources. 1980. *World Conservation Strategy: Living Resources Conservation for Sustainable Development* (Morges, Switzerland: IUCN), 55 pp.

4. Proceedings of the Second International Symposium on Coral Reefs. 1974. *The Great Barrier Reef Committee*, Brisbane, Qld., Australia, 2 vols, 1,383 pp.; Proc. Third International Coral Reef Symposium, Rosenstiel School of Marine and Atmospheric Research, Univ. of Miami, 2 vols, 1,284 pp.

5. A. Domm. A Review of Selected Recreational and Professional Activities on the Great Barrier Reef, unpublished report to the Great Barrier Reef Marine Park Authority.

6. E. Frankel. 1978. *Bibliography of the Great Barrier Reef Province* (Canberra: Australian Government Publishing Service), 204 pp.

7. Domm. 1977. A Review.

8. MSJ Keys Young, Planners Pty. Ltd. 1977. Public Participation Methods and Program, Unpublished report to GBRMPA.

9. D.F. Gartside. 1986. "Recreational Fishing," Proc. National Conference on

Coastal Management, Coffs Harbour, October 1986, New South Wales Pollution Control Commission, Sydney, New South Wales, Australia.

10. D. Hopley. 1980. "Gazetteer and Classification of the Reefs of the Great Barrier Reef," Unpublished Report to GBRMPA.

11. D. Warne. 1979. "Application of LANDSAT to Management of the Great Barrier Reef," Workshop on the Northern Sector of the Great Barrier Reef, GBRMPA, Townsville, pp.388–402.

12. Great Barrier Reef Marine Park Authority. 1983. Research Report 1976/82, Townsville, Qld., Great Barrier Reef Marine Park Authority, 133 pp.

13. J. McEvoy and T. Dietz, eds. 1977. *Handbook for Environmental Planning* (New York: John Wiley), 325 pp.

14. Domm. 1977. A Review.

15. Great Barrier Reef Marine Park Authority. 1978. Workshop on the Northern Sector of the Great Barrier Reef, April 1978, Townsville, 462 pp.

16. Great Barrier Reef Committee. 1979. Conservation and Use of the Capricorn and Bunker Groups of Islands and Coral Reefs, Report for GBRMPA, Brisbane, Great Barrier Reef Committee, 42 pp.

17. Enviroment Science and Services. 1979. Zoning Strategy Study Based on the Proposed Capricornia Section of the Great Barrier Reef Marine Park. Report to the Great Barrier Reef Marine Park Authority by Environment Science and Services in association with the Zoning Strategy Study Group, 285 pp.

18. Ibid.

Chapter 12

1. A.H. Robinson. 1976. *Recreation, Interpretation and Enviromental Education in Marine Parks*, Proc. International Conference on Marine Parks and Reserves, Tokyo, 12–14 May 1975, IUCN Publications New Series 37:99–119.

2. J. Baldwin, 1989. Great Barrier Reef Marine Park and Queensland Marine Park, Mackay/Capricorn Section Basis for Zoning, GBRMPA, Townsville, Qld., 58 pp.

Chapter 13

1. D.F. Gartside. 1986. *Recreational Fishing*, Proc. national Conference on Coastal management, Coffs Harbour, October 1986, New South Wales Pollution Control Commission, Sydney, New South Wales, Australia.

2. A.M. Kay and M.J. Liddle. 1985. Manual for the Assessment, Location and Design of Reef Walking Activities, Unpublished technical report to GBRMPA, 32 pp. New Ref. 1988.

3. J.C. Halas. 1985. *A Unique Mooring System for Reef Management in the Key Largo National Marine Sanctuary*, Proc. 5th. Internat. Coral Reef Symp. 4:237–42.

4. Great Barrier Reef Marine Park Authority. 1981. The Nomination of the Great Barrier Reef by Commonwealth of Australia for Inclusion on the World

Heritage List. GBRMPA, Townsville, Australia, 37 pp.

5. G. Claridge. 1987. "Assessing Development Proposals." in @UNESCO Coral Reef Management Handbook, 2nd ed. 1987, eds R.A. Kenchington and B.E.T. Hudson (Jakarta:UNESCO), pp. 131–38.

6. Great Barrier Reef Marine Park Authority. 1988. Cairns Zoning Plan Review: Issues, Great Barrier Reef Marine Park Authority, Townsville, Qld., Australia, 24 pp.

Chapter 14

1. B.C. Epler. 1987. "Whales, Whalers and Tortoises," *Oceanus* 30 (2):86–92.

2. C. Darwin. 1845. *Journal of Researches into the Natural History and Geology of the Countries Visited during the Voyage of H.M.S.* Beagle *Round the World Under the Command of Cpt. FitzRoy* (London: R.N. John Murray).

3. C. Darwin. 1859. *On the Origin of Species by Means of Natural Selection, or, The Preservation of Favoured Species in the Struggle for Life* (London: John Murray).

4. A. Laurie. 1984. "Marine Iguanas: The Aftermath of El Nino", *Noticias de Galapagos* 40:9–11.

5. J. M. Broadus and A.G.Gaines. 1987. "Coastal and Marine Area Management in the Galapagos Islands," *Coastal Management* 15:75–88.

6. F. Arcos, F. Cepeda, T. Rodriguez, and J. Villa. 1988. *Plan de Zonification de la Reserva de Recursos de Galapagos* (Quito, Ecuador: Ministry of Agriculture), 65 pp. + 51 pp. appendices.

7. R.A. Kenchington. 1989. "Tourism in the Galapagos Islands: The Dilemma of Conservation," *Environ. Cons.* 16(3):227,232,236.

8. R.S. DeGroot. 1983. "Toursim and Conservation in the Galapagos Islands," *Biol. Cons.* 26(4):291–300.

9. Arcos, et . al. 1988. *Plan de Zonification.*

10. De Groot, 1983. "Tourism and Conservation."

11. Arcos et. al. 1988. *Plan de Zonification.*

12. G.M. Wellington. 1975. The Galapagos Costal and Marine Enviroments: A Report to the Department of National Parks and Wildlife, Quito, Educador; 1976a. "A Prospectus: Proposals for the Management of a Galapagos Marine Park," *Noticias de Galapagos* 24:9–13; 1976b. "Suggestions for the Management of a Galapagos Marine Park," *Noticias de Galapagos* 25:5–12.

13. Broadus, Gaines. 1987. "Coastal and Marine Area Management."

14. Ibid.

15 Ibid.

16. R.A. Kenchington. 1989. "Planning for the Galapagos Marine Resources Reserve," *Ocean and Shoreline Management* 12:47–59.

17. P. Ryan, ed. 1987. "The Galapagos Marine Reserve," *Oceanus* 30(2):104.

18. A.G. Gaines and A. Mareano, eds. 1987. Scientific Research and the Galapagos Marine Reources Reserve, draft proceedings of a workshop in Guayaquil

Ecuador, April 1987, INOCAR, Guayaquil, Educador, 273 pp.

19. G.M. Wellington. 1984. "Marine Environment and Protection," in *Key Environment Series: Galapagos Islands*, eds. J.E. Treherne and R. Perry (Oxford: Pergamon), 247–67.

20. G.G. Kelleher and R.A. Kenchington. 1982. "Australia's Great Barrier Reef Marine Park: Making Development Compatible with Conservation," *Ambio* 11 (5):262–67.

21. Great Barrier Reef Marine Park Authority. 1986. Annual Report 1985/86, Townsville, Qld., Australia, 106 pp.

22. Great Barrier Reef Marine Park Authority. 1987. Zoning the Southern Sections (Capricorn and Capricornia Sections), Townsville, Qld., Australia, 61 pp.

Chapter 15

1. Ministry of Planning and Development. 1987. *Statistical Yearbook of Maldives* (Maldives: Male), 180 pp.

2. R. Sathiendrakumar and C. Tisdell. 1989. "Tourism an the Economic Development of the Maldives," *Ann. Trsm. Res.* 16:254–69.

3. R. Sathiendrakumar and C. Tisdell. 1986. "Fishery Resources and Policies in the Maldives," *Marine Policy* 4:279–93.

4. Sathiendrakumar and Tisdell, 1989. "Tourism.". "Tourism and the Development of the Maldives," *Massey Jnl. of Asian and Pacific Business* 1(1):27–34.

5. P.G. Flood. 1974. *Sand Movements on Heron Island—A Vegetated Sand Cay, Great Barrier Reef Province Australia,* Proc. 2nd Internat. Coral Reef Symp., pp. 387–94; 1977. "Coral Cays of the Capricorn and Bunker Groups, Great Barrier Reef Province, Australia, *Atoll Res. Bull.* 195: 1–7.

6. Ibid.

7. Sathiendrakumar and Tisdell, 1986. "Fishery Resources."

8. F. Berkes. 1985. "Fishermen and the Tragedy of the Commons," *Enviromental Conservation*, 12(3): 199–206.

9. S. Wells and A. Edwards. 1989. "Gone with the Waves," *New Scientist* [11 Nov. 1989], pp. 29–33; C. Woodroofe. 1989. Maldives and Sea-Level Rise: An Environmental Perspective, Report on Visit to the Republic of Maldives, Feb. 1989, Dept. of Geography, Univ. of Woolongong, N.S.W., Australia, 64 pp.

10. R.A. Kenchington. 1985. Report on Missions to the Republic of Maldives, October, 1983 and February 1985, unpublished report to the Ministry of Planning and Development, Male, Republic of Maldives, 83 pp.

11. Wells and Edwards, 1989. "Gone with the Waves."

Chapter 16

1. K.A. Bekiashev and V.V. Serebriakov. 1981. *International Marine Organizations*, trans. V.V. Serebriakov (The Hauge: Martinus Nijhoff), 578 pp.

2. S. Mankaby, ed. 1984. *The International Maritime Organisation* (London: Croom Helm), 376 pp.

3. Ibid.

4. P. Hulm. 1983. *A Strategy for the Seas: The Regional Seas* Programme, *Past and Future* (Geneva: UNEP), 28 pp.

5. Institute of Sanitary Engineering, Milan. 1982. *Waste Discharge into the Marine Environment* (Milan: Polytechnic of Milan), 432 pp.

6. World Commission on Environmental Development. 1987. *Our Common Future* (Oxford: Oxford Univ. Press), 400 pp.

7. G.G. Kelleher and R.A. Kenchington. (in press) *Guidelines for Establishing Marine Protected Areas* (Morges, Switzerland: IUCN).

8. UNESCO. 1974. *"Task Force On Criteria and Guidelines for the Choice and Establishment of Biosphere Reserves"* (Paris:UNESCO), 61 pp.

9. M. Batisse, 1986. "Developing and Focusing the Biosphere Reserve Concept," *Nature and Resources* 20 (4): 1–12.

10. Ibid.

11. W. Gregg and B. McGean. 1985. "Biosphere Reserves: Their History and Promise," *Orion Nature Quarterly*, pp. 41–51.

12. G.R. Francis. 1988. *Biosphere Reserves in Developing Countries: The Canadian Experience* Proc. of the Man and Biosphere Technical Symp., Denver, World Wilderness Congress.

13. Batisse. 1986. "Developing and Focusing the Biosphere Reserve Concept."

Chapter 17

1. A.H. Robinson. 1976. *"Recreation, Interpretation and Environmental Education in Marine Parks,"*Proc. International Conference on Marine Parks and Reserves, Tokyo, 12–14 May 1975, IUCN Publications New Series 37, pp. 99–119.

2. Budowski, G. 1976. "Tourism and Environmental Conservation: Conflict, Coexistence of Symbiosis?" *Env. Cons.* 3 (1):27–31.

3. Salm, R.V. 1985. "Integrating Marine Conservation and Tourism." *Int. Jnl. Envtal. Studies* 25:229–38.

4. G.E. Machlis and D.L. Tichnell. 1985. *The State of the World's Parks* (Boulder CO: Westview Press), 131 pp.

Index

Aborigines. *See* Australian Aborigines
abyssal plain, 16, 22
Acanthaster planci, 43-44, 116-117, 122-123
access. *See also* zoning - plans
  GBR, 118
  prohibitions, 68-69
accountability, identified in legislation, 64
activities. *See* uses
activity monitoring, 66, 100-101, 107-108
Addu Atoll, 187, 193, 196, 197
administration. *See also* financial arrangements; government agencies; interagency relations; management
  GBR Marine Park Authority, 135-172
  legislative aspects, 64-64
  Maldives, 191
  personnel, 64, 95-98
  structure, 151
Administrative Appeals Tribunal, 161
Administrative Decisions Judicial Review Act (Aust), 161
advertising, 167
advocacy. *See also* interest groups marine scientists, 1-2
aerial photography. *See* remote sensing
agencies. *See* government agencies; interest groups
aid programs, 177-178, 191-204, 219
aircraft
  access to Maldives, 187-188
  dumping protocols, 210
  GBR regulations, 134
alienation
  defined, 40
  economics, 214
  identifying and analyzing impacts, 45
  Maldives marine habitats, 193, 194, 196, 197
  management solutions, 214-215
  threat to environments, 2, 37, 41, 46
amenity
  Cairns Section, 170-171
  defined, 41
  management goals, 42, 49-52
  Mediterranean Regional Sea, 211
  perceptions, 149, 215,-216
  valuing, 49-51
Amoco Cadiz (ship), 11, 206
anchorages, 155
antifouling paints and agents, 47
appeals, 161
Area Statements, 168
Ari Atoll, 200, 201
artisanal fisheries. *See* subsistence fisheries
attachment of personnel, 97-98
Australia
  Administrative Appeals Tribunal, 161
  High Court. *See* High Court of Australia
  jurisdiction over GBR, 127-129
  legislation, 127-134. *See also* specific legislation eg *Marine Parks Act* (Qld)
  Minister for the Environment responsibilities, 165-166
  politics, 119-125
Australian Aborigines, 115
Australian Conservation Foundation, 119
Australian Institute of Marine Science, 117
Australian Littoral Society (Queensland), 119, 120
Australian Survey Office, 141-142
authority, defined by legislation, 63

Bangladesh, 20
Bass Strait, 121
bibliographies. *See* documentation
biogeography
  Galapagos Archipelago, 173
  GBR, 114
biological diversity, 114
biological resources. *See also* fisheries
  exploitation, threat to environments, 2
biology
  Galapagos Archipelago, 173
  Great Barrier Reef, 112-114

Biosphere Reserve Program, 211-213, 221
biota. *See also* species
effects of pollutants, 46-49
Black Sea, 209 boundaries
delineated by legislation, 63, 130
jurisdictional issues, 15-23
Brundlandt Commission, 13
Buonoparte's Gulf, 121

Cairns Section, review of zoning plans, 170-171
Capricorn/Bunker area. *See also* Capricornia Section;
Capricorn Ridge
encompassed by Capricornia Section, 146
guano deposits, 116
Capricornia Section, 144, 146. *See also* Mackay/Capricorn Section
Capricorn Ridge, 143-144. *See also* Capricorn/Bunker area
Carribbean Regional Sea, 209
cartography, 141-142, 200
causeway construction, Maldives, 196, 197
CDF. *See* Charles Darwin Foundation for the Galapagos Islands
Central Region of Great Barrier Reef, 11, 112
Charles Darwin Research Station, 174
charting, 141-142, 200
chlorine, effects on biota, 49
civil liability, conventions, 207
coastal zone management. *See* land based ac tivities
coastlines
engineering, 2, 41, 446, 167, 193-194, 196, 314
jurisdictional issues, 16-24
COBSEA. *See* Coordinating Body on the Seas of East Asia
Cockburn Reef, 112
collecting. *See* fisheries
commons (concept of ownership), 7-8, 10, 42, 142
fisheries, 45, 46, 78, 214
Commonwealth of Australia. *See* Australia
community leaders, 87
compensation rights
conventions, 207
recognized in legislation, 66
compliance. *See also* enforcement; surveillance
defined by legislation, 63
conferences and seminars, 97, 177-178
Conferences of Plenipotentiaries of the Coastal States of the Mediterranean Region for the Protection of the Mediterranean Seas, 209-210
conflict resolution. *See also* multiple use; resource allocation
single issues, 10, 43
subsistence fisheries, 7-8
connectivity. *See* linkage
consensus. *See* public opinion
conservation. *See also* preservation; sustain ability
conventions, 20
general issues, 44-52
interest groups, 119, 137, 144, 158-159
international cooperation, 12-13
public interest, 11, 119
recognized in *GBR Marine Park Act*, 133
relationship with tourism, 217-218
constitutional issues. *See also* jurisdiction; sovereignty
GBR Marine Park, 126-134
constraint analysis, 89-90
construction materials. *See also* coastlines - engineering
Maldives, 188, 193, 196
consultants, 140, 177-178, 192, 219
consultation. *See also* public participation; social surveys
part of planning program, 75, 76
prior to plan drafting, 85-89
tactics, 77-78, 91, 92
consumers. *See* users
contiguous zone, 20-21, 25
continental shelf
conventions, 20, 22
defined, 22
legislation, 123, 127
Continental Shelf (Living Natural Resources) Act 1968, (Aust), 123, 127
continental slope, defined, 22
controls, available to managers, 68-74, 161-168
Convention for the Protection of the Mediterranean Sea against Pollution, 210
Convention on fishing and the Conservation of Li ing Resources of the High Seas, 20
Convention on the Continental Shelf, 20, 126-127
Convention on the High Seas, 20
Convention on the Prevention of Marine Pollution by Dumping Wastes and Other Matters, 207, 208
Convention on the Territorial Sea and the Contiguous Zone, 20, 22
conventions, 14-27, 42, 207-208, 210. *See also* international cooperation; specific conventions eg *Convention on the High Seas*
Cook, James (explorer), 115
Coordinating Body on the Seas of East

Asia, 98, 220
Coral Reel Management Handbook (UNESCO), 75
coral reefs and coral islands. *See also* Maldives; specific reefs and islands eg Great Barrier Reef
  assessment of development proposals, 104-106, 166
  crown of thorns phenomenon, 43-44, 116-117, 122-123
  dynamics, 127, 194-195
  jurisdictional issues, 127-128
  mining, 46, 120, 195-196
  public education approaches, 99
  traditional management, 8
  zoning goals, 77
costing. *See* valuing
COTSAC. *See* Crown of Thorns Starfish Advisory Committee
counseling, 101
courses. *See* conferences and seminars
crown of thorns starfish, 43-44, 116-117, 122-123
Crown of Thorns Starfish Advisory Committee, 142
cultures
  Australian Aborigines, 115
  Maldives, 191
  recognized in legislation, 64, 67
  resource allocation, 7-8
  subsistence traditions, 8
  valuing, 3
currents
  affecting water masses, 29
  pollution transport, 9

Darwin, Charles (naturalist), 116, 173
data analysis. *See* information gathering and analysis
data collection. *See* monitoring; surveillance
declaration, 1-2, 132
deep seabed, 16, 22
defense closure areas, 154, 160
degradation. *See* impacts
developing nations. *See also* Ecuador; Maldives
  international aid, 219-220
  training, 96
development
  impacts, monitored, 73, 103
  politics, 118-125
  proposals, procedures, 104-106, 166-167
  risks to environment, 217
  tourism, stages, 216
disasters, 11, 121, 137, 206
  conventions, 207, 210
dispersants, effects on biota, 49
diversity, 114
documentation, 44-45, 78. *See also* publications
  Capricorn Ridge, 144
  GBR Marine Park, 140
draft plans. *See also* draft zoning plans
  preparation, 76, 85-91
draft zoning plans
  Galapagos Marine Resources Reserve, 176-182
  GBR, 153-156, 157
dredging, 46
dumping, conventions, 207, 208, 210
dynamics of ecosystems, 14-15, 28-29, 43, 102-103
dynamite fishing, 45

East African Regional Sea, 209
East Asian Regional Sea, 209. *See also* Coordinating Body on the Seas of East Asia
economics
  development
    assessment, 218
    increasing public interest in conservation, 119
  Queensland, 118-119, 124
  environmental management, 2
  Galapagos Archipelago, 174
  Maldives, 184-190
ecosystems. *See also* terrestrial ecosystems
  characteristics affecting management, 28-39
  dynamics, 28-29, 30, 43, 102-103
  scales exceeding jurisdictions, 14-15
Ecuador
  economy, 174, 181-182
  Presidential Decree 1810-A, 176
  support for protection of Galapagos Archipelago, 173-174
Ecuadorian Foundation for the Conservation of Nature, 182
education. *See also* public education; training
  importance in management programs, 94
  increasing expectations of environment, 215
  interest groups, 10
  legislative recognition, 67
  for users of management areas, 98-100
EEZ, 22, 24, 25
effluents. *See also* rivers
  effects on biota, 48
EEZ, 22, 186
Ellison Reef, 120
Emerald Agreement, 139
emergencies. *See* disasters
endangered species, 32. *See also* extinction; sites - of special conservation significance

Endeavour River, 115
Endeavour (ship), 115
energy dynamics, 29-39, 31
enforcement. *See also* self enforcement
  Galapagos Marine Resources Reserve, 180
  management program key element, 94
  personnel, 96
  powers and duties provided in legislation, 67
engineering. *See* coastlines - engineering
environmental impact statements, 73, 166, 167
Environment Council (Maldives), 199
Environment Protection (Impact of Proposals) Act (Aust), 165
environments
  degradation. *See* impacts
  management. *See* management
  plans. *See* plans
  threats. *See* impacts
equipment restrictions, 71
erosion, Maldives, 194-195, 197
exclusive economic zone, 22, 24, 25
exclusive fishing zone, 22, 186
exploitation. *See* uses
exploration. *See* also shipping
  history, 6-7, 115-116, 173
  mineral. *See* petroleum - drilling
external activities. *See also* land based activ ities
  controlled by *GBR Marine Park Act*, 136, 138-139
  controlled by regulations, 67
  recognized in legislation, 63
extinction, 33, 41. *See also* endangered species
extraction, 40, 46. *See also* dredging; mining

facilities prohibitions, zoning plans, 169, 171-172
familiarization tours, 96-97
federal-state relations, 119-125
  Emerald Agreement, 139
  jurisdictional issues, 16, 26-27
  permit procedures, 166
  policy development for GBR Marine Park, 135-139
  protection of GBR, 124
  sensitivity over international conventions, 206
feedback public participation programs, 149
fertilisers, effects on biota, 48
financial arrangements
  Galapagos Marine Resources Reserve, 181-182
  identified in legislation, 66
fisheries. *See also* specific fishing methods eg pulse fishing; subsistence fisheries
  conventions, 20
  economics, 3-4, 9, 45-46, 51
  exploitation trends, 9, 214
  Galapagos Archipelago, 174
  GBR, 114, 116, 123
  impacts, 40
  interest groups, 10, 158-159
  legislation and regulation, 9, 12, 42
  limitations on, 71
  Maldives, 186, 187, 188, 196-198
  opportunity threat profiles, 57-58
  zoning goals, 77, 155
fishers
  attitudes, 9, 12, 79, 141
  sources of information, 82-83
  target for public education, 99
fish poisons, 45
Flinders, Matthew (navigator), 116
Fly River, 112
Food and Agriculture Organization, 3
food preservation, 9
foreign fishing, 123
freshwater. *See also* rivers
  effects on biota, 48
  lens, 192, 194
Fundacion Ecuatoriana para la Conservacion del la Naturaleza. *See* Ecuadorian Foundation for the Conservation of Nature
Fundacion Natura. *See* Ecuadorian Foundation for the Conservation of Nature

Galapagos Archipelago, 173-183, 208
Galapagos Marine Resources Reserve, 173-183, 181
Galapagos National Park, 174, 175, 180
Galapagos National Parks Service, 174, 177, 180
Ganges Delta, 19
Gan Island, 187-188
garbage, 207
GBRMPA. *See* Great Barrier Reef Marine Park Authority genetic mixing, 30, 32
genetic reserves. *See* preservation sites
geographic information systems, 90
geographic scale. *See* scale
geography
  Galapagos Archipelago, 173-175
  GBR, 109-112, 110, 111, 112, 113
  Maldives, 184
  Queensland, 114
geomorphology
  Great Barrier Reef, 112
  Maldives, 194, 195
GESAMP. *See* Group of Experts in Sampling

Marine Pollution
GIS, 90
global action. *See* international cooperation
GMRR. *See* Galapagos Marine Resources Reserve
goals
  of environmental management, 60-61
  key to plan development, 76-78
  outlined in legislation, 61
government agencies
  creation minimized, 63
  education of personnel, 94, 96-98, 99, 220-221
  involvement in plan drafting, 67, 86-88
  marine activity involvement, 54-55
  responsibilities identified in legislation, 64
Great Barrier Reef, 109-172
  crown of thorns phenomenon, 43-44
  similar issues to Galapagos Archipelago, 174
Great Barrier Reef Committee, 1116, 120, 44
Great Barrier Reef Consultative Committee, 132, 144, 157
Great Barrier Reef Marine Park, 12, 118-712, 131
  planning methods, 75-93
  sections. *See* sections of GBR Marine Park
Great Barrier Reef Marine Park Act 1975, provisions recognized in Reef Use Plans, 169
Great Barrier Reef Marine Park Act 1975 (Aust), 130-134
  involving Queensland in decisions, 129
  section 31, 145
  section 32(7), 140, 144, 145
  section 38, 136-139, 145
  section 38, 136-139, 145
  section 66(2)(e), 136, 138-139
Great Barrier Reef Marine Park Authority
  established under *GBR Marine Park Act*, 130-132
  establishment phase, 135-146
membership of International Commission (Ecuador), 177
  procedures, 147-172
greenhouse effect, 199
Green Island, 116, 138
Group of Experts in Sampling Marine Pollution, 11
guano, 116

harvesting. *See* fisheries
Hawaii, 20
hazardous chemicals, 207, 210
health issues, Maldives, 193
heavy metals, 42, 49
herbicides, effects on biota, 46
heritage, valuing, 3-4
Heritage Program (UNESCO), 178. *See also* World Heritage Areas
Heron Island, 117, 138, 143
High Court of Australia, 127, 138
high seas
  conventions, 20, 207
  defined, 22
  protected areas, 24
history
  Galapagos Archipelago, 173-174
  GBR, 115-117
  seas, 6-13
human activities. *See* uses
human population
  Galapagos Archipelago, 174
  GBR coastal zone, 114
  Maldives, 184
  related to impact, 40, 161, 163
hydrocarbons, effects on biota, 47-48

Iceland, 20
image. *See* public relations
IMCO. *See* International Maritime Co-ordinating Organization
IMO. *See* International Maritime Organization
impacts. *See also* specific topics eg fisheries - impacts
  assessment, 217-218. *See also* monitoring
    GBR, 118, 140, 145
    importance, 94
    petroleum drilling, 120-121
    practical approaches, 103-106
    procedures, 166
  categorized, 40
  conservation perspectives, 50-52
  costing, 3
  increase, 7-10, 43
  integration, 53-59
  land based activities, 9, 40, 214
  legislation, 165-166
  minimization, 2, 39, 40, 42-44
  new or altered, 43, 94, 103-106, 172
  protected areas, 218-219
  public awareness, 10, 11
  transport across jurisdictional boundaries, 2
implementation. *See* plans - implementation
Indonesian Marine National Parks system, 12
induction briefings, 96
information gathering and analysis. *See also* documentation; monitoring; surveillance
  difficulties, 35-37
  Galapagos Marine Resources Reserve, 178
  procedures
    permit applications, 164-168

procedures (*continued*)
planning, 148
premanagement phase, 78-83, 140-142
zoning plan preparation, 151-152
infrastructure support, 63, 180-182
infringements. *See* compliance
INGALA. *See* National Institute for the Galapagos
INOCAR. *See* Naval Oceanographic Institute (Ecuador)
Instituto Nacional de Pesca. *See* National fisheries Institute (Ecuador)
Instituto Nacional Galapagos. *See* National Institute for the Galapagos
Instituto Oceanografico de la Armada. *See* Naval Oceanographic Institute (Ecuador)
interagency relations, 3. *See also* federal-state relations; international cooperation
for effective conservation management, 15, 52-56
encouraged under Biosphere Reserve Program, 212
Galapagos Archipelago, 176-177, 180-182
recognized in legislation, 63-63
interest groups. *See also* specific groups eg Australian Conservation Foundation; specific interest areas eg conservation - interest groups; users
attitude to multiple use management, 12
attitude to public participation, 148
involvement in planning, 50, 85-8867, 149, 158-159
lobbying, 125
local and nonlocal, 149
practical approaches to education, 98-100
reasons for growth, 10, 119
role identified in legislation, 64-65
inter-institutional Commission of Government Ministers (Ecuador), 176, 177, 178, 180
internal waters, defined, 21
international Association for Biological Oceanography, 220
International Commission (Ecuador), 177
International Conference on Marine Parks and Protected Areas (1975; Tokyo), 11
International Convention for the Prevention of Marine Pollution from Ships (MAR POL), 207, 208
International Convention for the Prevention of Pollution of the Sea by Oil, 207, 208
International Convention on Civil Liability for Oil Pollution Casualties, 207
International Convention on the Establishment of an International Fund for Compensation for Oil Damage, 207
International Convention relating to Intervention on the High Seas in Cases of Oil Pollution Damage, 207
international cooperation, 2, 3, 4-5, 11-13, 205-213, 219. *See also* international law
Galapagos Marine Resources Reserve, 177-179, 182
Maldives, 191-192
International Coral Reef Symposia, 140
international law. *See also* conventions
relevant to environmental management, 24-26, 205-213
International Maritime Co-ordinating Organization, 11, 206-208. See also International Maritime Organization
International Maritime Organization, 3, 206-208. *See also* International Maritime Co-ordinating Organization
International Oceanographic Commission, 220
International Sea-Bed Authority, 22, 24
International Training and Advice Program (NOAA), 178
Natural Resources, 11, 211, 220
General Assembly (17th), 13
international waters, 20, 22, 24, 207
interprofessional relations, 3, 12, 212
intertidal communities, adapted for living in tidelands, 31
intra fauces terrae, 21
investigation. *See* information gathering and analysis
IOC. *See* International Oceanographic Com mission
islands. *See also* coral reefs and coral islands; specific islands eg Heron Island
Galapagos Archipelago, 173-183
jurisdictional issues, 15, 127-128
Maldives, 184-204
IUCN. *See* International Union for the Con servation of Nature and Natural Resources

James Cook University of North Queensland, 117, 141
Jukes (explorer), 116
jurisdictional issues. *See also* sovereignty
complicating resource management, 10
GBR, 124, 127-129, 130, 138, 146, 165-166
legal background, 14-27
transport of impacts across boundaries, 2

Kuwait Regional Sea, 209

Lady Musgrave Island, 169
land based activities. *See also* coastal zone management; external activities

land based activities (*continued*)
economics, 3, 214
focus of Regional Seas initiatives, 26
impacts, 9, 40, 214
Mediterranean Regional Sea, 210
planning and development, 53
recognized in legislation, 63
land/sea interface, 15, 21, 127-128, 138
lava flows, 20
law of the sea. *See* United Nations Conferences on the Law of the Sea (UNCLOS)
legislation. *See also* specific topics eg resources - legislation
activity-specific, 55-56
guidelines, 61-67
lobbying by interest groups, 125
liability, conventions, 207
licenses, 72
life cycles
influencing management needs, 70
strategies, 33-35, 34. *See also* reproductive strategies
limestone mining, 118, 120, 195
limitations, 69-72, 74
linkage
demanding international cooperation, 14-15, 205
factor affecting management, 29-30, 37, 39, 42, 52, 142, 145
between Galapagos Islands and sea, 173, 175
local impacts lessened, 41
preventing evaluation of management controls, 33
recognized in legislation, 62
with terrestrial environment, 9, 40, 63, 214
literature. *See* documentation
littering, Maldives, 193
Lizard Island, 117
lobbyists. *See* interest groups
low water mark
definitional conflicts, 128
jurisdictional issues, 17

MAB. *See* Man and Biosphere Program
Mackay/Capricorn Section. *See also* Capricornia Section
special management areas, 160-161
zoning plan, 161, 162
Malaysian Marine Protected Area Program, 12
Maldive fish, 184-186
Maldives, 184-204, 185
Male Atoll, 184, 188, 192, 193, 196, 201
management. *See also* conservation; economic management;
impacts - minimization; pollution - control; resource allocation
administration, 2-3
areas. *See also* protected areas
boundaries, 14-27
coordination, 3, 10, 12, 52-53. *See also* interagency relations; international cooperation; multiple use
establishing a framework, 60-75
Galapagos Marine Resources Reserve, 180-182
GBR, 147-172
impact minimization, 2, 39, 40, 42-44
implementation. *See* plans - implementation
international arrangements, 24-26, 42, 205-213
longterm commitment, 2, 218-219
Maldives, 199-204
methods, 42-59
multidisciplinary approaches, 3, 4, 212
personnel. *See* managers
planning. *See* planning
plans. *See* plans
programs, 94-108. *See also* plans - implementation
significance of scale, 2, 28, 33-35, 37, 39, 219
sovereignty issues, 14-27, 124, 138, 206
strategies. *See* planning
sustainability as goal, 3, 60, 214, 219
trends, 214-221
unique characteristics of marine ecosystems, 28-39
managers
interactions with scientists, 78
training, 95-98, 220-221
Man and Biosphere Program
mangroves, 8, 19, 41, 45
manufacturing, Maldives, 190
mapping, GBR, 141-142
mariculture. *See* fisheries
marine. *See* term without 'marine' eg marine parks. *See* parks
Marine Parks Act (Qld), 169
MARPOL. *See International Convention for the Prevention of Marine Pollution from Ships*
Mediterranean Regional Sea, 26, 209-211
migrating birds, 70
mineral resources, sovereignty issues, 14
mining. *See also* limestone mining
exploration. *See also* petroleum - drilling
industry views on *GBR Marine Park*, 136, 144
prohibition by *GBR Marine Park* Act, 133, 136-139
monitoring

monitoring (*continued*)
activity, 66, 100-101, 107-108
alternative to permit application, 167
management program key element, 94
practical approaches, 101-103
role in evaluating plan outcome, 107-108
specified in legislation, 66
user pays, 167
Montego Bay Convention, 16, 20, 21
motivation. *See* public support
Mourilyan Harbour, 116
multidisciplinary approaches, 3, 4, 212
multiple use. *See also* resource allocation
applications of permit systems, 73
approach to planning, 12
attitude of interest groups, 12
complicating impact analysis, 45
Galapagos Marine Resources Reserve, 178-179
GBR, 12, 118, 133, 145, 158-172
Maldives, 201-204
management methods, 50, 50-59
planning techniques, 89-91
recognized in legislation, 62
valuing components, 82
zoning plans as a management approach, 158-172
muro ami, 45
Murray Islands Group, 109

National fisheries Institute (Ecuador), 177
National Institute for the Galapagos, 177, 180
National Oceanic and Atmospheric Administration (US), 177, 178, 179
national parks
acceptable use, 168
adjacent to marine areas, 138, 155
Ecuador, 174, 175, 178, 180, 181-182
extended to marine environments, 12
legislation, 12, 169
National Parks and Wildlife Act (Qld), 169
National Science Foundation (US), 177
Naval Oceanographic Institute (Ecuador), 177
navigation, regulation, 134
nitrogen, effects on biota, 48
NOAA. *See* National Oceanic and Atmospheric Administration (US)
noncompliance. *See* compliance
nongovernment agencies. *See* interest groups
nonrenewable resources, 14
Northern Region of Great Barrier Reef, 112, 113
North Sea, 121
notice of intent to prepare a plan, 153
notification, alternative to permit application, 167
NSF. *See* National Science Foundation US)

objectives. *See* goals
Oceanic Grandeur (ship), 121
oceans. *See* seas
Oceanus (journal), 177
oils. *See* petroleum
One Tree Island, 117, 143
opinion surveys, 87, 140
opportunity threat profiles, 56-58
Orpheus Island, 117
outcome evaluation
of management controls, 33
of management plans, 107-108
ownership. *See also* commons (concept of ownership);
sovereignty
existing rights recognized in legislation, 64

Pacific Ocean, opportunity threat profiles, 56-68
paper parks, 219
Papua New Guinea, 20
parks. *See also* national parks; specific parks eg Great Barrier Reef Marine Park
legislation, 61-67, 130-134, 169
management. *See* management
on paper only, 219
Parliamentary procedures, zoning plan acceptance, 167
patrols, 100-101
personnel. *See* enforcement - personnel
penalties, provided by legislation, 67
perceptions, 10-11, 37-38, 49-52, 149, 215-216
permits, 73
allowing ongoing evaluation of impacts, 103
procedures for GBR Marine Park, 161-168
personnel
administrative, 64, 95-98
enforcement, 96
management, 78, 95-98, 220-221
political, 96
research, 67
pesticides, effects on biota, 47
petroleum
drilling
disasters, 121, 137, 206
impact assessment, 120-121
legislation, 127
preclude by GBR Marine Park Act, 136-139

petroleum (*continued*)
program established for GBR, 118
hydrocarbons, effects on biota, 47
petroleum (*continued*)
oil annex to MARPOL convention, 207
pollution conventions, 207, 208, 210
spills, 22, 121, 137, 206
Petroleum (Submerged Lands) Act 1967 Aust), 127
phosphates. *See* fertilisers; sewage; sewage-detergent phosphates
physical extraction. *See* extraction
planning
catering for multiple use, 50, 50-59
complexity, 42-44
controls, 68-74
delays, 78
Galapagos Marine Resources Reserve, 178-180
GBR, 133, 147-172
guidelines, 90
implementation trends, 218-221
programs, 75-76
public participation legislated, 65
supporting documents, 83-85
plans. *See also* draft plans; zoning - plans
development
Galapagos Archipelago, 177-180
GBR, 147-172
Maldives, 192-199
finalization, 76, 92-93
implementation, 94-108. *See also* management - programs
new and altered uses, 103
Regional Seas, 209
review, 66, 94-95, 107-108
specified in legislation, 65
policy
development for GBR, 135-139
outlined in legislation, 61
politics
awareness education for personnel, 96
national, 119-125. *See also* federal-state relations
public demanding action on degradation, 11
state, 118-125. *See also* federal-state relations
pollution. *See also* specific pollutants eg herbicides; wastes - disposal
control, 2, 11
conventions, 42, 161, 207, 208, 210
economics, 82, 214
GBR, 118, 145
impacts, 2, 40, 41-42, 46-49
Maldives, 192
Mediterranean Regional Sea, 210-211
pollution (*continued*)
transport, 9
Poritidae, 114
powers. *See also* jurisdiction
powers (*continued*)
adequately defined by legislation, 63, 67
precedence, 63, 137-138
preliminary environmental reports, 166
premanagement phase, procedures, 78-83
preservation sites, 159, 212-213. *See also* sites - of special conversation significance
Princess Charlotte Bay, 115
process impacts, type of impact, 41-42
proclamation, 1-2, 132
productive capacity. *See also* sustainability
maintenance a goal, 60
prohibitions, 68-69, 169, 171-172
project management, 151
prosecutions, 101
protected areas. *See also* parks; preservation sites
classification, 12
creation, trends, 11
generic threats, 218-219
goal of environmental management, 2, 60
impacts from external activities, 63
interest groups, 10
international cooperation, 11, 24, 26
legislation, 12, 63
Mediterranean Regional Sea protocols, 26
opportunity threat profiles, 58-59
public awareness of need for, 11
purposes of use established by plans, 73-74
Protocol concerning Co-operation in Combating Pollution of the Mediterranean Sea by Oil and other Harmful Substances in Cases of Emergency, 210
Protocol Concerning Mediterranean Specially Protected Areas, 26
Protocol for the Prevention of Pollution of the Mediterranean Sea by Dumping from Ships and Aircraft, 210
Protocol relating to Intervention on the High Seas in cases of Marine Pollution by Substances other than Oil, 207
protocols. *See* conventions
publications
design, 88-89, 91, 148, 152
draft zoning plans, 156-157
training, 98, 220-221
zoning plans, 157
public awareness. *See also* education
of degradation, 2, 10, 11
Maldives, 198
practical approaches, 98-100
priority in management, 215

public education. *See* education
public opinion, 149-150. *See also* public participation; public perceptions; social surveys
public participation. *See also* consultation
established in legislation, 65, 67
following plan drafting, 91-92, 156-157
Galapagos Marine Resources Reserve, 179
part of planning program, 75, 76
prior to plan drafting, 85-89
program design, 141
provided for in *GBR Marine Park Act*, 133
role in planning *GBR Marine Park*, 147-150
role in zoning plan preparation, 153, 156-157
public perceptions, 10-11, 37-38, 49-52, 149, 215-216
public relations, 44, 99, 101
pulse fishing, 46

Queensland
Department of the Co-ordinator General, 142
geography, 114
involvement in GBR management legislated, 129, 132
jurisdiction over GBR, 127-129
legislation, 166, 169
Mining Warden, 120
Minister for Mines, 120
politics, 118-125
Queensland Marine Parks, 165
Queensland National Parks and Wildlife Service, 169
Queensland State Environment Assessment legislation, 166
questionnaires, 88-89
quotas, type of limitation strategy, 71-72

Raine Island, 116
Ramsar Convention, 70
'reasonable use', 90, 140, 144, 145
reclamation. *See* alienation
records management, in public participation programs, 149-150
recreational use. *See also* tourism
competing for resources, 10, 215-218
increasing on GBR, 114
interest groups, 10, 158-159
opportunity threat profile, 58
quotas, 71-72
technological advances, 10
valuing, 3-4, 82
zoning considerations, 77, 156
recruitment, 95
Red Sea/Gulf of Aden Regional Sea, 209
reefs. *See* coral reefs and coral islands
Reef Use Plans, 168-169
reference sites. *See* preservation sites
referenda, 86-87
refuges. *See* preservation sites
Regional Seas Program, 24-26, 208-211
areas proposed, 26
developing conventions and protocols, 3
training programs, 98, 220-221
regulations. *See also* fisheries - legislation and regulation
authority provided in legislation, 66-67
inconsistent with federal law, 134
Maldives, 201
power to make under *GBR Marine Park Act*, 133-134
remote sensing, 36, 100, 141-142, 200
renewable resources. *See also* biological resources
conventions, 20
legislation, 123, 127
representative environment. *See* protected areas
reproductive strategies, 31-33
Republic of Ecuador, 173-174, 176, 181-182
Republic of Maldives, 184-204
research. *See also* information gathering and analysis; preservation sites
difficulties, 33, 35-37, 79
field stations, 116-117, 174
GBR, 115, 116-117
GBR Marine Park establishment phase, 142
interest groups, 10
literature. *See* documentation
permit applications for, 164
personnel, 67. *See also* scientists; technical specialists
priorities, 57-59, 79-83
reason for legislative delay, 65-66, 78
recommended for Maldives, 200
specified in legislation, 66
to support resource exploitation, 8
techniques, 78
valuing, 3-4
zoning considerations, 155
resource allocation. *See also* conflict resolution; multiple use
fisheries, 214
goal of management, 2, 60
licenses, type of limitation strategy, 72
Maldives, 204
management approaches on GBR, 158-172
traditional management, 7-8
trends, 10-11

resources. *See also* nonrenewable resources; renewable resources
exploitation
causing process impacts, 41-42
sovereignty issues, 14-15
trends, 6-10
legislation, 7-8, 61-67, 130-134
sea a source of, 6-10
responsibility, identified in legislation, 64
review literature. *See* documentation
review processes, specified in legislation, 66
right of innocent passage, 21, 207
rivers. *See also* effluents; freshwater
influencing tidal reference points, 19
transferring impacts, 9, 214
Royal Commissions on Petroleum Exploration and Production, 121, 126
Royal Society of London, 115, 116

salinity, effects on biota, 48
San Cristobal Island, 179
sanctuaries. *See* preservation sites
Santa Barbara, 121
Santa Cruz Island, 174, 179
satellite remote sensing. *See* remote sensing
Save the Reef campaign, 123
scale
demands international cooperation, 14-15, 219
factor in management, 2, 28-29, 37, 39
of GBR Marine Park, 118
of government decision making, 14
of life cycle, 33-35
Scientific Committee for Oceanic Research, 220
scientific literature. *See* documentation
scientists. *See also* research - personnel
interaction with managers, 78
role, 1, 220
SCUBA diving, 35-36, 72
seabed, 16, 22
sea/land interface. *See* tidelands
sea level rise, Maldives, 198-199
seas, history, 6-13
Seas and Submerged Lands Act 1973 (Aust), 127, 129, 130, 138
seawater, as an environment, 30-31
secondment, 97, 178
sections of GBR Marine Park. *See also* specific sections eg Capricornia Section
extent, 145-146
sedimentation
changing tidal reference points, 19-20
effects on biota, 47
land based activities causing, 9
self enforcement, 67, 101
seminars. *See* conferences and seminars

Servicio Parque Nacional Galapagos. *See* Galapagos National Parks Service-
sewage
annex to MARPOL convention, 207
detergent phosphates, 48
effects on biota, 48
shipping. *See also* exploration
conventions, 42, 207, 208, 210
disasters, 11, 121, 206
GBR, 115-116, 154
hazards, 7
history, 6-7, 8-9
interest groups, 10
shipwrecks, 155
simplicity, a virtue, 74, 89, 167
single issue management, 10, 43
sites
biologically isolated, 30
dependence, unusual in marine ecosystems, 32-33
GBR zoning strategy study, 143
of special conservation significance. *See also* preservation sites
fewer in marine environments, 32-33, 41
legislative aspects, 70
Mediterranean Regional Sea, 210
protected by international conventions, 208
zoning guideline, 154-155
specific guidance on preferred development. *See* Area Statements; Reef Use Plans; special management areas
site-specific management strategies, 34
skill licenses, type of limitation strategy, 72
skills transfer. *See* training
skipjack tuna, 184-186
social surveys, 87, 140
South East Asian Marine Parks, 98
South East Pacific Regional Sea, 209
Southern Region of Great Barrier Reef, 109-11, 111
South West Atlantic Regional Sea, 209
South West Pacific Regional Sea, 209
sovereignty. *See also* jurisdiction
ceded under international conventions, 206
federal-state conflict over GBR, 124, 138
legal background, 14-27
spatial controls
in GBR Marine Park, 159-161
type of limitation strategy, 69-70
special interest groups. *See* interest groups
special management areas, 160-161. *See also* preservation sites
species. *See* biota; endangered species
SPNG. *See* Galapagos National Parks Service

staff. *See* administration - personnel
storms. *See* winds
structural impacts, 40-41
subsistence fisheries
  Australian Aborigines, 115
  collapse, 45
  conflict resolution, 7-8
  valuing, 82
  zoning goals, 77
subtidal areas
  communities, 31
  jurisdiction, 127-128
surfactants, effects on biota, 49
surveillance
  practical approaches, 100-101
  role of evaluating plan outcome, 107-108
  specified in legislation, 66
surveys. *See also* social surveys
  design, 88-89
  reason for legislative delay, 65-66, 78
sustainability. *See also* conservation; productive capacity
  approach to planning, 12
  goal of management, 3, 60, 214, 219
  legislation, 62, 64
  of tourism operations, 217
Swains Group, 109

tactics, in dealings with the public, 77-78, 91, 92
Taiwan, 123
Technical Commission (Ecuador), 177, 178-179
technical specialists, participation in planning, 67
technical support, international aid, 177-179, 182, 191-192, 199, 220
technological advances
  hazards of shipping reduced, 7, 8-9
  human impact on seas increased, 7-10
  increasing human impact, 215-216
tectonic phenomena, changing tidal reference points, 20
telephone communications, 148
temperature, adverse effects on biota, 48-49
temporal controls, type of limitation strategy, 70
terminology, standardized in legislation, 64
terrestrial ecosystems, comparison with marine ecosystems, 28-39
territorial waters, 20, 21, 22, 23, 24, 25. *See also* 3-mile territorial sea
threats. *See* impacts
3-mile territorial sea, 14, 20, 21, 23, 24, 26-27, 126-127, 138
tidelands. *See also* intertidal communities
  defined, 21
tidelands (*continued*)
  jurisdiction, 15, 127-128
  unprotected by *GBR Marine Park Act*, 138
tides, 16-17. *See also* low water mark
  affecting water masses, 29
  pollution transport, 9
  reference points, 17-20, 18, 19, 129. *See also* low water mark
time controls, type of limitation strategy, 70
Torres Strait, 109, 112, 115
Torrey Canyon (ship), 11, 121, 206
tourism. *See also* recreational use
  education for operators, 99
  Galapagos Archipelago, 175, 176, 181-182
  GBR, 114, 118, 124, 158-159, 170-171
  interest groups, 158-159
  Maldives, 187-190, 193
  Mediterranean Regional Sea, 211
  opportunity threat profile, 58
  quotas, 71-72
  relationship with conservation, 217-218
  technological advances, 10
  use of marine environments, 10, 215-218
  valuing, 3-4
  zoning plans, 77
trade, 7, 115, 116
traditions. *See* cultures
tragedy of the commons. *See* commons (concept of ownership
training, 94, 95-98, 220-221
transport. *See* aircraft; shipping
trawl fishing, 123
treaties. *See* international cooperation tropical regions, resource management, 8
turbidity, effects on biota, 47
12-mile territorial sea, 20, 21, 23, 24
200-mile territorial sea, 20, 22, 23, 24

UNCLOS. *See* United Nationals Conferences on the Law of the Sea
UNEP. *See* United Nations Environment Program
UNESCO. *See* United Nations Educational, scientific and Cultural Organization
UNESCO Coral Reef Management Handbook, 75
United Nations, establishing International Maritime Co-ordinating Organization, 206
United Nations Conference on the Law of the Sea UNCLOS), 15
  conventions, 20-24
  (1st: 1958: Geneva), 16, 20
  (3rd: 1982), 16, 20-24, 129
United Nations Educational, Scientific and Cultural Organization, 173, 191-192,

United Nations Educational, Scientific and Cultural Organization (*continued*) 211, 220, 221
United Nations Environment Program, 220
  Regional Seas Program. *See* Regional Seas Program
United States Environment Protection Administrative Procedures, 104-106, 166
University of Rhode Island, 177
URI. *See* University of Rhode Island
'user pays' principle, 167
users. *See also* interest groups; multiple use
  consultation with, 83, 85-88
  contributions to planning GBR Marine Park, 147-148
  direct compared with indirect, 53-56
  local compared with nonlocal, 149
  planning submissions, 149
  practical approaches to education, 98-100
uses. *See also* external activities; land based activities
  defining meaning of 'reasonable', 140-141
  history, 6-8
  impact on GBR, 140
  objective assessment criteria, 145
  prohibitions, 69
  requiring permits in Mackay/Capricorn Section, 163

valuing
  amenity, 49-51
  components of a multiple use area, 82
  difficulties, 79
  fisheries, 51
  high quality environments, 3-4, 215
  impact, 2-3
  use of natural waters, 214
volunteers, 96

wastes
  conventions, 207, 208
  disposal, 4, 6, 82, 214. *See also* pollution
West and Central African Regional Sea, 209
Wheeler Reef, 128
WHOI. *See* Woods Hole Oceanographic Institution
Wildlife Preservation Society of Queensland, 119, 120
winds
  affecting water masses, 29
  changing tidal reference points, 17, 19, 19
Woods Hole Oceanographic Institution, 178, 179
workshops. *See* conferences and seminars
World Commission on Sustainable Development, 211
World Conservation Strategy, 62, 140
World Conservation Strategy, 12-13
World Heritage Areas, 165, 173-174
World Wilderness Congress (4th), 13, 211

zoning
  integrating incompatible goals, 76-77
  plans. *See also* Area Statements; draft zoning plans; Reef Use Plans
    confused with zoning strategy study, 144
    Galapagos Archipelago, 179-180
    GBR, 147-172
    legislation, 133-134
    Maldives, 201, 204
    Male Atoll (Maldives), 202-203
    provisions for new and altered uses, 103
    spatial controls on activities, 70
    specified in legislation, 65
  strategy study, 143-144
  town and land use approaches, 142-142

About the Author

Richard Kenchington is Secretary to the Coastal Zone inquiry of the Resource Assessment Commission of Australia. For many years he was responsible for planning and policy development for the Great Barrier Reef Marine Park Authority. He is a member of the Commission for Sustainable Development of the World Conservation Union.

He has served as a consultant to international agencies and governments on conservation, management and sustainable use of marine and coastal environments. He has been a Marine Policy Fellow at the Woods Hole Oceanographic Institution. His writings include articles in Environmental Conservation, Biological Conservation, Ocean and Shoreline Management and two editions of *The Coral Reef Management Handbook*, published by UNESCO.

F